The Dawn of Death

The Dawn of Death

The Bible and Our Mortal Bodies

Simon Howard

WIPF *&* STOCK · Eugene, Oregon

THE DAWN OF DEATH
The Bible and Our Mortal Bodies

Wipf & Stock
An Imprint of Wipf and Stock Publishers
199 W. 8th Ave., Suite 3
Eugene, OR 97401

www.wipfandstock.com

PAPERBACK ISBN: 978-1-4982-9922-0
HARDCOVER ISBN: 978-1-4982-9924-4
EBOOK ISBN: 978-1-4982-9923-7

Manufactured in the U.S.A.

Dedicated to Kati.

Till death us do part.

. . . that only Tree
Of Knowledge, planted by the Tree of Life;
So near grows Death to Life, whate'er Death is—
Some dreadful thing no doubt . . .

—John Milton, *Paradise Lost*

Contents

1

Why Do We Die?

"MOST PEOPLE WOULD RATHER die than think; in fact, they do."[1]
So the philosopher Bertrand Russell described a prevalent reluctance to apply a measure of rationality to our brief experience of the world. This reluctance is especially prevalent when it comes to thinking about death. To be sure, an awareness of the prospect of death drives much, if not most, of how we live: whether our efforts are directed towards denying, delaying, or accepting death, or even bringing it on, we are arguably preoccupied with the subject. But this is more the subject of living and dying than of death and decay. The preoccupation only serves to prove Russell's point that people are so caught up in immediate experience and imminent prospect that few spare time for critical reflection on the immovable facts of existence that limit our experience and determine our prospects. "Our days are numbered, so let us get on with living and perhaps preparing to die before the count is up." But why need the number of our days be finite? "The clock is ticking, so let us make the most of every moment." But why does the clock slow and always stop? The most fundamental questions rarely get asked while we live in their shadow. "Why do we die?" is one of the most fundamental of all.

There are two competing answers to this question, each based on contrasting value judgments about death. One holds physical death to be a fate profoundly unfitting for beings that, among

---

1. Bertrand Russell, source unknown, common attribution.

other remarkable abilities, can ask questions about their own existence—such as why they die. On this view, we should not have to
die—physical death is an intrusion—and we have not always had
to die. The other answer sees the human being as, however distinctive in some respects, one among the many living organisms that
inhabit this planet—organisms that, without exception, physically
expire as part of a recurring cycle of life. On this view, we are subject to the same contingencies as every other form of life, and so a
finite physical lifespan is, and always has been, our lot.

"We Are Not Meant to Die."

The first of the two perspectives just mentioned arises from a keen
conception of each human life as invested with meaning—meaning that is expressed in relationship with others and confirmed by
awareness of self. This meaning is differently nuanced for each individual—"I am unique"—and it has an inherent permanence. Herein
lies the problematic nature of human physical death: existential
meaning is physically embodied, and the dissolution of the bodily
vehicle of meaning presents a challenge to the continuance of meaning. It takes only a moment's reflection on the demise of any person
to be arrested by the scandal of the fact that, to all appearances, they
are no more. (It takes no reflection at all when the person was someone we knew well.) "The death of any man diminishes [us]," in John
Donne's words,[2] not just because "no man is an island" but because
it is the baffling negation of a unique constellation of attempts and
achievements, hopes and dreams, pleasures and preoccupations,
memories—and meaning. The strength of this sentiment is stronger
the more brief the life lost, but it is by no means absent when faced
with the demise of someone who has lived the usual "threescore
years and ten" (Psalm 90:10 KJV). The strange obsolescence of our
physical frame brings to nought a life spent trying, with some suc-
cess, to transcend our physical limits.

2. John Donne (1572–1631), *Meditation* XVII.

Faced with this finitude in contradiction to our felt permanence, some of us relativize our physical lifespan with the thought of another mode of living beyond our physical death. Bound up with this expectation is a degree of dissatisfaction with our present existence: we expect a life beyond death to be a better one than this. Intuition and experience tell many that there is something defective about the material world of which we are presently part, and physical death, as the consequence and cause of so much that we feel should not be, is something that should not happen—and will not in the life hoped for. For those who believe in a good Creator God, no fault in the material world can be attributed to the Creator. Our physical mortality, if it is one fault among many, must be contrary to the Creator's perfect intention, an ungodly intrusion on human existence. Since to our knowledge none of those first physically immortal persons has lived to tell the tale, the intrusion must have occurred very early indeed in human history—within the foreshortened lifetime of the earliest individuals.

In the absence of any means of probing the distant past for answers to matters such as these, the early chapters of Genesis can easily serve to confirm a hunch born of present experience. Those chapters contain an account of God's giving life to a first couple, his forbidding them from an action on penalty of death, their disobedience nevertheless, and God's pronouncement of a punishment on the couple and their progeny containing an allusion to the ending of their physical lives. This account is but the opening scene in a long saga of now-mortal generation after generation, until the notable death of an individual whose ancestry is traced to the very first man.[3] It is notable not least because that individual did not remain physically dead but came back to unending bodily life as the first of many more who will do the same. Thus a supposed original immortality is restored, the riddle of human physical death resolved, and the meaning of life recovered.

---

3. Luke 3:23–28.

"How Could We Not Die?"

There are, however, some facts of material existence that sit uncomfortably with the scheme just laid out, facts which are all the more difficult to ignore since the scriptural account on which the scheme is based bears witness to them. Human beings are one among many kinds of living creatures,[4] and humans, like other forms of physical life, sustain their material selves by eating,[5] and "multiply and fill the earth" by reproducing.[6] However much the human creature is distinguished by alone bearing God's image (whatever that is), its physical frame resembles that of other living beings in its formation, composition, properties, and eventual fate. This clear homology between human and beast has not been wholly lost on humanity through history, despite presuming for most of its existence that the universe revolves, not just around the planet on which it lives, but around the consciousness of self it feels. Nevertheless, it took until just 150 years ago for a man who did think before he died to articulate, based on more extensive observation of nature than most, not just the physical homology between living beings of such numerous kinds, but a theory of homogeny—of common biological descent.

In putting forward his theory,[7] Charles Darwin did not particularly concern himself with the physical existence of his own human race, but he was not unaware of the implications it would have, if correct, for an understanding of human being. At any rate, his opponents soon sharply brought the implications to his attention: that humans are one of a kind and, moreover, they are kin with all other forms of physical life; with regard to death, humans die because animals (and all other organisms) do. And animals *must* die, so the claim goes, because only by doing so can the strongest of the species—especially those younger, fitter members—prosper in surroundings that are subject to constant change. On

4. Genesis 1:26.
5. Genesis 1:29–30.
6. Genesis 1:28.
7. Darwin, *Origin of Species* (1859).

this view, the limitation on our physical lifespan is a legacy of our long formative phase in a branching tree of diverse biological life.

This presents no uncertain challenge to a simple reading of the Genesis narrative in which the first human individual is described as having been brought into being "from the dust of the ground"[8] by the direct intervention of his Creator when the universe was just six days old, six thousand years ago. The length of time needed for Darwin's scheme to work itself out from microbe to man is vastly greater. But the trickle of systematic observation of biological life and death before Darwin soon became a rising torrent of "evidence" that, if not providing conclusive proof of his theory of "the origin of species," has at least rendered it plausible and established it as a working hypothesis of unsurpassed explanatory power in biology. Much of this support takes the form of fossilized remains of dead creatures, preserved deep in layers of sedimentary rock laid down in an orderly and gradual manner. Among this is a small but significant collection of material with striking similarity to our modern-day human form. There are striking differences too, so they are better described as hominin, not human, remains. But the morphological resemblance to our physical selves is even closer than that of now-living primates, so the remains serve as a putative biological link between us and some of the jungle's present inhabitants. However uncertain the interpretation of this evidence, one thing is sure: the creatures from which these remains derive are now dead. If the physical origins of our own race lie in them, then either our earthly lives have also always been inherently bounded by physical death or we are such a stupendous exception that the biological resemblance between us and other forms of life is of little significance.

Of course, we *are* a stupendous exception in our extraordinary (if underused) ability to think before we die, including our ability to think about death. But does this make us a biological exception too? In a way, it does. By technological innovation we can, in some measure, control our environment rather than it controlling us; our surpassing ingenuity enables us—not nature—to be, to some extent, the natural selector. Despite this, though, our lives are still strangely

8. Genesis 2:7.

circumscribed by physical death. We can increase our longevity by a decade or two, and we can "immortalize" cells in the laboratory, but our physical mortality presents an obstacle seemingly impossible to avoid. Since Darwin's day, biology has become less a voyage to catalogue life on earth in its multifarious forms and more an exploration of the workings of the basic unit of life, the cell. Perhaps there, in the genetic code, the biological clock can be located, and a way found to keep it ticking. Or is the way to the tree of unending physical life so well guarded that it is impassable even by science?

Towards an Answer of Sorts

We set out by stating that there are two competing answers to the question "Why do we die?" and we have outlined them both. The first starts out from a deeply felt existential significance, which, because it is felt so deeply, claims an inherent permanence: a permanence that, strangely and painfully, does not extend to its physical embodiment, but should—and would in a perfect world. This perspective finds authorization in no less than the word of God himself—or an interpretation of it—which gives an account of the Creator's breathing life into human bodily form but our first forebears' bringing death onto themselves and us. The second answer also starts from a commonly observed fact of human existence: that of our sharing with other living creatures a closely similar physical form and a certain biological fate. The view that human physical mortality is a natural necessity has developed by recourse to science, not Scripture. At one level, scientific observation can reveal no more than is already obvious: we die, without exception. But at another level, when that observation takes place within a theoretical framework in which physical death is a prerequisite for continually renewed, improved physical life, the only permissible inference is that, not only do we physically die, but we have always done so and things could not possibly be otherwise.

These two answers compete as two sides of a heated debate. But the thinking Christian cannot allow him/herself to be drawn easily towards one or the other, for s/he experiences within him/

herself the appeal of both: s/he is more than an animal, yet not unlike one; s/he exercises control of his/her physical body, yet is bounded by it; s/he feels him/herself somehow immortal, yet is physically not. Treading a careful line between each opposing sentiment and school of thought, a Christian consideration of the puzzle of human physical mortality (which this book aims to be) must start out with a clear idea of which aspects are open to investigation and which are not. Three aspects are within our reach, as follows. Interpretations of Scripture are used to buttress one side in the debate, so we will (i) closely examine the relevant passages to assess whether they do indeed portray humankind as having once lost a physical immortality which it originally possessed. Whatever the result of that, we will also (ii) consider whether any biblical material could be employed in support of the other perspective, depicting human physical death as equivalent to that of other living beings. In considering the content of Scripture on a subject that is also a matter for science, another aspect must be addressed: (iii) the relative reliability of the pronouncements of Scripture and science on natural phenomena, including the phenomenon of human physical death.

At least three aspects of the puzzle are not open to investigation. It is not just the length of our physical lives that is limited, but so is our ability to answer all our deepest questions. In this book we will not (i) be able to solve the "mind-body problem," a problem that, as is evident from our foregoing discussion, is relevant to our topic but may be intractable. Neither will we (ii) be able to prove or disprove an ape-to-human transition, a transition that, because it is no longer observable, we can know about in only a qualified sense. Our ability to know about ourselves being restricted, our knowledge of God is even less: rather obviously, it is not possible (iii) to discern the Creator's original intention regarding the duration of human physical embodiment; we cannot peer into the mind of God to discover whether our physical lifespan is limited by definition or defect. In venturing to think about death before we die, any conclusions we draw should be marked by a circumspection that befits our mortal condition.

2

Scientific Perspectives
on Physical Mortality

IN THE FIRST CHAPTER, we started on our way towards fathoming why we die by sketching out two contrasting positions. This next chapter will look more closely at one of them: the idea that humans have always been physically mortal. In particular, we will survey the science behind it. Subsequent chapters will focus on the other position, that mortality is something our bodies acquired at some time in the past.

The theory of evolution is talked about more often with regard to biological *life* than death. This is only to be expected, since it was formulated as an explanation of the present diversity of living forms on earth and continues to serve as the majority explanation for their morphological similarities and differences. However, the ending of biological lives is as central to the evolutionary process as the generation of new forms of life. "Once multicellular carbon-based life forms began to exist, then a dynamic natural order in which life *and death* are integral parts became an inevitable consequence."[1] This is so for two reasons.

First, in Darwin's own words, there is a "struggle for existence"[2] between individuals of the same species dependent on resources that are limited relative to the reproductive potential of

1. Alexander, *Matrix*, 352, emphasis added.
2. The title of chapter 3 in Darwin's *Origin*.

the species. Against a background of variation of phenotype within the species, some individuals lose out in the struggle: they die. Over periods of environmental change, those individuals with inherited variations favouring survival in the new conditions will be "naturally selected"; the process leaves in its wake dead, less-well-adapted individuals or even, when the change is great enough, an extinct species.

Second, in addition to death being a feature of evolution by "survival of [only] the fittest," as just described, the living world largely sustains itself by being "red in tooth and claw."[3] In other words, most species of multicellular carbon-based organisms (plants excepted) derive their energy from other, usually simpler, carbon-based organisms situated earlier in the food chain. Both herbivorism and carnivorism are "structural necessities"[4] for the living world as it has evolved and as it exists today.

It should be clear from the previous two paragraphs that, according to biological science, without death there can be no evolution of complex life and indeed no animal life at all. It is not immediately clear, however, what bearing this has on human physical death. Generally people do not die in competition with their neighbors over scarce resources, and humans are second to none in all their food chains. But we still eventually die.

In this book, our interest is in human death by growing old: age-related death, the fact of a finite lifespan. "Human physical mortality" best labels our present concern, if we exclude from the notion of "mortality" susceptibility to death by accidental injury.[5] Despite first impressions, science does in fact speak to

---

3. Note that neither "survival of the fittest" nor "nature, red in tooth and claw" are phrases coined by Darwin. The first comes from Herbert Spencer (*Principles of Biology* [1864]) and the second from Alfred Tennyson (*In Memoriam A.H.H.* [1850]).

4. Peacocke, *Science*, 136.

5. Death by accidental injury is not a concern in this book because, given that the world is governed by laws of physics, it is easy to understand scientifically why soft-bodied creatures such as ourselves can die from physical injury. It is less easy to understand why, if we can physically thrive for any time at all, our lifespan is not indefinite.

this phenomenon as well as to the other kinds of biological dying mentioned above.

First, there is a growing amount of evidence that human beings, like all other species of living things, are products of evolution at least in their physical form. The evidence indicates that we are the result of a long process of natural selection for those organisms that were our ancestors and against those that are not. (Though those that are not our ancestors lived and live in ecological niches different from our own.) The relevance of this to understanding human physical mortality is that it therefore becomes possible to consider the human species sharing with at least some other species the same physical basis of a limited lifespan. What this basis is is a second relevant area of scientific investigation. It involves research into the proximal cause(s) of aging and dying—the genetic and physiological mechanisms—in a range of species, as well as consideration at a more theoretical level of the distal cause(s), in particular whether age-related mortality is a corollary of the evolutionary process itself.

Since in this book we aim to engage seriously with the scientific perspective that human physical mortality is and always has been a biological necessity, we will next review the scientific findings and thinking that lie behind this perspective.

The Origins of *Homo Sapiens*

The evolutionary scheme for the history of life is complex, but crudely it runs as follows: after the appearance of a single common ancestor of all subsequent forms of life between four and six billion years ago, single-celled prokaryotes became established, then multicellular fungi and early plants, followed by animals. Fish were the first animals, from which amphibians, then reptiles and, from reptiles, mammals and birds developed. It is beyond our scope here to present the evidence and arguments for this textbook scheme. Of most interest in connection with biological death is evidence for an animal-to-human transition such that humans are, biologically, animals. In discerning such a transition, the starting point is our close anatomical and genetic similarity to the modern chimpanzee.

On an evolutionary reckoning, the time required for the genetic divergence between chimps and humans to reach its current extent is six to eight million years.[6] It can be inferred, therefore, that present-day chimps and humans share a last common ancestor, now extinct, that lived this long ago. No fossil of this organism has been found. There have, however, been multiple discoveries of fossilized organisms belonging to both branches after the divergence. Those on the developmental line leading to modern humans—collectively termed "hominins"—can be recognised by their possession of anatomical features distinctive to us in contrast to the great apes of today: a dense, load-bearing neck to the femur and/or a forward position of the foramen magnum where the spine meets the skull (both indicating bipedalism), and/or smaller canine teeth, and premolars with two cusps.[7]

Fossils with the features just mentioned can be grouped into genera and species based on the range of variation seen within species of interbreeding individuals today. The precise divisions between these species are disputed, but there are just over twenty. The age of each fossil is determined by radioisotopic dating of either overlying rock or (at younger sites) the fossil itself. Roughly speaking, the twenty-plus species form a chronological series, from the oldest specimen, of *Sahelanthropus tchadensis*, six to seven million years old, to fossilized remains of our own species, *Homo sapiens*, which are just tens of thousands of years old.[8] This is not to say that there was a simple succession of species, one after another; the dating indicates that two or more species of hominin (e.g., *Homo erectus* and *Homo heidelbergensis*) coexisted, even in the same region, as did two genera (*Paranthropus* and *Homo*).

---

6. Lockwood, *Story*, 7.

7. Ibid., 10–12.

8. The oldest fossils of *H. sapiens* date to 150,000–200,000 years old (ibid., 103).

"Human evolution was not a single path towards modern humans but instead a more complex and diverse array of forms."[9]

The fact that the oldest hominin remains have all been found in eastern, central, and southern Africa indicates that continent as the location for evolution of the ancestor we share with the great apes. Discoveries of younger fossils in Asia and Europe reveal, however, several waves of migration out of Africa. *Homo erectus* probably first spread into Asia, then *H. heidelbergensis* into Europe (becoming, in the face of climatic change, *H. neanderthalensis*).[10] The findings suggest a later, twofold migration from Africa of *H. sapiens*, first into the Middle East (only to be replaced there by Neanderthals) and then more successfully into all continents except the Americas, supplanting other species of hominin.[11] There are indications that each of the four species just mentioned had abilities that we could subjectively call "human": respectively, the abilities to control fire, to hunt with a spear, to make simple adornments, and, at least in *H. sapiens*, to produce symbolic art.[12]

It would not be difficult to present the material evidence of human origins in much greater detail, but the sketch just given serves to convey the scientific perspective that our own species is in developmental continuity with non-human biological life. This provides a rationale for the scientific search for a common physical basis to the mortality of both ourselves and our non-human living relatives.

Understanding Mortality by Modeling Aging

It is not physical mortality itself that is the direct object of scientific investigation, but rather the physical process of aging, also known as senescence. The relation between age and mortality in humans can be stated simply: with increasing age an individual

9. Ibid., 52.
10. Ibid., 73, 93.
11. Ibid., 102–6.
12. Ibid., 73, 87, 94, 104.

is also increasingly likely to die. Mortality increases exponentially with age not simply due to the passage of time but because of *aging*, involving a loss of physiological function and an increase in susceptibility to degenerative disease.[13]

Even in the absence of disease before death, organ failure can be detected after death; "old age" per se is probably never the immediate cause of death. "The phenotype of human aging is one in which practically any system, tissue, or organ can fail. This indicates an intrinsic phenomenon affecting the whole organism and leading to the 'weakest link' failing, resulting in death."[14] The furthest aim of research into this phenomenon is to arrest the biological processes involved, so that average human lifespan may be extended, even indefinitely.

Humans share the aging phenotype just described with all other mammals.[15] In part because experimental investigation of aging is in its infancy, and in part because the lifespans of those mammals most closely related to humans are impractically long, mammals cannot serve as models for the study of aging. Instead, the models most commonly used are: (i) unicellular organisms (e.g., yeast), (ii) the roundworm, and (iii) the fruitfly.[16] Unfortunately, all these organisms have aging phenotypes different from mammals. Mice and cultured human cells are in principle more representative but, in practice, results from mice may be artifacts stemming from the use of laboratory strains, and human studies in vitro may poorly simulate in vivo processes.[17] Clearly, experimental research into biological aging is subject to huge methodological limitations and is carried out largely on the most basic assumption that "understanding how life itself works may help us understand aging."[18] Nevertheless, research on this assumption has borne

---

13. Magalhães, "What Is Aging?"

14. Ibid.

15. Magalhães, "Some Animals Age, Others May Not."

16. Magalhães, "Human Aging Model Systems."

17. Ibid.

18. Ibid.

verifiable results in other areas of biochemistry (e.g., metabolic pathways), so it holds promise in the science of aging too.

The experimental results gained so far indicate that there is a genetic component to aging and therefore to mortality. Retardation of senescence and increased longevity have been achieved by genetic manipulation in all the model organisms just mentioned.[19] But this merely confirms what was previously known; a genetic component to aging was already evident from the facts that, for example, (i) there are rare congenital diseases with the symptom of accelerated aging, and (ii) identical twins die on average 36 months apart in comparison to 106 months for other siblings.[20] The key debate surrounds the relative contributions of genetic and non-genetic (e.g., environmental) factors in aging. Environmental damage in the sense of "wear and tear" is of little significance in aging because all organisms display a capacity for repair. Studies on laboratory models suggest, however, that "damage" in the form of the influence of physiological factors (e.g., metabolic products) may modulate an underlying genetic "clock." Presently, understanding the overlap between genetic and non-genetic inputs is hindered by the difficulty of distinguishing between causes and effects.

The Questions Science Can, and Cannot Yet, Answer

While laboratory research focuses on the cellular mechanisms, it does little to advance our knowledge of the underlying reasons for aging: why it takes place at all, and why longevity has an upper limit. The theoretical, not experimental, biology of aging has more to say about these. It is not possible here to present more than a few theories of aging in brief, giving only an impression of their general character.[21]

---

19. Magalhães, "Grandparents of Tomorrow."

20. Clark, *Sex*, 82.

21. See Magalhães, "Evolutionary Theory of Aging" for a full review.

The "disposable soma theory" starts out from the observation that age-related death occurs only in multicellular, sexually reproducing organisms, not unicellular asexual ones. On this account, mortality is a trade-off for the species-survival benefits of sexual reproduction. With the advent of the distinction between somatic and germ cells, the former acquired obsolescence.

The point of departure for another theory, pleiotropy, is the increasing statistical probability of death by accidental injury with increasing age, even discounting physiological aging. The age distribution of the population is thus skewed towards younger individuals. On an evolutionary reckoning, natural selection favors genes beneficial at earlier ages but will not disfavor other, deleterious genes expressed at a later age. This theory has some experimental support from the increased longevity of fruit flies achieved by artificial selection of older, over younger, individuals.

The descriptions just given conceal some formidable difficulties in each of these theories. As in all scientific inquiry, progress towards discovery will be made by allowing the theoretical and the experimental to inform each other. On the question of age-related physical mortality, scientists have not reached the point of discovery; so far the science is merely descriptive (and poorly so), providing an insufficient basis for the predictive assertion that without age-related mortality there can be no higher, sexually reproducing forms of life.[22] Biologists are on surer ground with the more modest claim that our own species, through its development from hominin predecessors and, before them, ancient apes, has never been other than physically mortal. Further, humans share the biological basis (as yet obscure) for this with at least some of the planet's non-human inhabitants.

22. So Clark, *Sex*, 76.

3

Biblical and Scientific Statements about Facts of Nature

THE QUESTION OF WHETHER or not human beings have always been inherently physically mortal is a question of natural history. Even if there is a theological aspect to the question, concerning God's creative intention or original human potentiality, physical mortality (or immortality) is (or perhaps was) a property of our existence as part of the material world—"nature." Since at least the nineteenth century, empirical science has been the principal way of studying nature and of uncovering the length and course of its history. In the last chapter, we surveyed the results of scientific study of physical mortality. While most Westerners today would hold that knowledge of nature can be gained by no other means than science, some—including some practicing scientists—maintain that there is another source of knowledge of natural historical fact, apart from nature itself: the Bible.

Beliefs about the origin of human physical mortality through most of Christian history and in much of the present-day church have been formed by recourse to Scripture. Theophilus of Antioch (AD 115–85), Irenaeus (died ca. AD 202), Aquinas (1225–74), Melanchthon (1497–1560), and Calvin (1509–64) all held the view that physical mortality became established in the natural world subsequent to, not at, a moment of creation, mostly by appeal to Genesis 2–3. Today, conservative biblical scholars, though generally reticent on the matter, invariably take the traditional line.[1]

Since the Bible is, on the broadest consensus, a book about God, and since the God it depicts is creator of all that has ever been (e.g., Genesis 1:1) and revealer of matters otherwise unknown (e.g., Daniel 2:22), it is not wholly unreasonable to regard the Bible as a place to turn for knowledge of nature. However, two assumptions behind this view warrant examination: first, that statements in the Bible about nature are as factually correct as those about any other aspect of reality; second (with respect to our topic of interest), that the Bible contains pronouncements about the origin of human physical mortality that are clear in their meaning.

The second of these assumptions will occupy us most, because the interpretive task demands meticulous care.[2] But the first assumption needs addressing more urgently, because it concerns an issue that must be settled before one turns to the text: the scientific (in)errancy of Scripture. Whether one starts from the position that the biblical text can only be correct on a question of natural fact or, alternatively, from a position that it may err, will influence whether we consult the Bible at all, and if we do, how we handle it. This, therefore, is our focus for the moment.

Literalism or Concordism

The view that the Bible is a sourcebook for knowledge of nature as well as of God is a particular extension of the belief that it is a book not just *about* God by also in some sense *by* God. While not everyone who holds to divine authorship of Scripture would extend it so far, the starting point for those who do is an affirmation of the character of the divine author. Thus the *Chicago Statement on Biblical Inerrancy* (1978) opens by declaring that God "is Himself Truth and speaks truth only," then states "that the written Word in its entirety is revelation given by God," and eventually arrives at a

1. For a survey of such scholarship, see Duce, "Comment," 161–63.
2. See chapters 4–6.

denial "that Biblical infallibility and inerrancy are . . . exclusive of the fields of history and science."[3] It is beyond our scope here to explore this doctrine of Scripture or to expound our own, although notions such as "revelation in entirety" and "given by God" clearly beg close scrutiny.

More pertinent here is the fact that those who affirm the Bible's inerrancy in "the field of science" must reckon with the findings of science itself in its own field of inquiry, the natural world. These findings do not always sit easily alongside Scripture. When faced with a scientific account of natural phenomena seemingly at variance with a biblical account, the inerrantist can take one of two routes: s/he can either assert that the biblical text, taken literally, fixes the structure and functioning of nature, regardless of the conclusions of science; or s/he can attempt to bring the biblical text into concord with the scientific account. We have yet to inspect the biblical material relevant to the question of the origin of human physical mortality, so we are not, at this early stage of our investigation, in a position to critique the literalist and concordist standpoints on this, our issue of particular interest. Rather, in this chapter we will deal with them at the general level of their interpretive approach.

Inerrantists' Interpretive Errors

Literalism and concordism have in common an anachronistic reading of Scripture, supposing the uses of language in texts that predate modern science by millennia to be comparable to linguistic usage in modern scientific reports: realist, informational, and explanatory. Such a reading arises in part from the assumption that the meaning of a divinely inspired text must be obvious.[4] And, to a modernist consciousness dominated by scientific and historical interests, statements of objective fact are the most obvious of all in

3. *Chicago Statement on Biblical Inerrancy*, articles III and XII.
4. Ramm, *View*, 45.

their meaning.[5] But whether stating facts of nature was among the biblical authors' intentions can only be gauged from exegesis of their texts; their purposes in writing may well have lain elsewhere. When they do allude to nature, as is unavoidable in any prolonged literary reflection on existence, their depictions, by dint of the period in which they wrote, are most likely to be pre-scientific: phenomenalist (according to immediate appearances) and non-postulational (not formulating theory or law).[6]

In addition to sharing this "linguistic confusion,"[7] literalists and concordists err in separate ways, primarily in their contrasting stances towards conventional science. Literalists exercise extreme suspicion of science insofar as it pronounces on matters on which God has already supposedly spoken in Scripture. Thus radiometric dating is flawed and evolutionary reasoning corrupt, not on technical or theoretical grounds, but in the face of the biblical statement that "the heavens and the earth were completed in all their vast array"[8] in six literal days. This is to abstain from serious engagement with the methods, evidence, and inferences of science, an enterprise that, though not entirely free of subjective interference, strives to minimize that interference by being self-critical and therefore self-correcting.

The self-correcting character of empirical science is a strength but also a weakness; science yields knowledge of nature which is the surest available but which is never more than provisional, always open to being revised, updated, or even overturned. Failure to recognize this is the concordist's misjudgement. By striving to bring the biblical text into agreement with the latest scientific approximations, s/he gives more credence to those approximations than they merit. "To place the biblical texts . . . in this kind of [scientific] arena is to . . . place them at the beck and call of the latest empirical evidence and interpretation."[9] The classic example of this

---

5. Hyers, *Meaning*, 28.
6. Ramm, *View*, 46–48.
7. Hyers, *Meaning*, 16.
8. Genesis 2:1 NIV.
9. Hyers, *Meaning*, 88.

fallacious approach—the overturning by Galileo's astronomical observations of a literal interpretation of certain Psalms, wed as it was to the received geocentric wisdom of Ptolemy—has worn tired with the telling, but other, more contemporary examples could be given.[10] Regarding our present topic of interest, the scientific viewpoint that human physical mortality is a legacy of our common descent from non-human species is unlikely to change, given that all the material evidence supports this view. But the more apparently secure the scientific position, the more obliged the concordist is to bring Scripture into conformity with it, and the more ingenious the interpretive feat must be if the biblical form of words does not explicitly endorse the science.

Behind the concordist endeavor, and arguably behind the literalist one too, is the laudable assumption that, on a doctrine of creation, Christian belief and the facts of nature are complementary, not contradictory. With this all Christians could agree, so long as they consider their faith rational and the world coherent. Concordists, however, take an illegitimate step further by asserting, on a mistaken doctrine of inerrancy, that the *Bible* and science are complementary too. We contend that it is not science and the Bible that are complementary, but science and *theology*. To compare the Bible and science is to compare two entities that are quite dissimilar. The Bible serves as a source for the theological commitments of Christian belief. Science is not itself a source but a set of testable interpretations of its source, empirical observation of nature.

Furthermore, it is not by any means a foregone conclusion that biblical statements about details of nature contribute to specifically *theological* formulations. It is only insofar as those formulations contain implications for understanding the material world of the past, present, or future that they intersect with the concerns of science and that they are complementary with correct scientific conclusions. But at the level of a dubious comparison between the

10. Assertions since the 1830s of a harmony between Genesis 1 and mainstream geology and biology (the 'Day-Age theory'). The assumption, dating from the 1940s, that the Big Bang theory accords with Genesis 1:1.

Bible and science, the purposes of the biblical authors, their uses of language, and the cosmologies of their cultures all preclude a complementarity that is more than accidental between their literary productions and modern scientific assertions. Even Francis Bacon, the early pioneer of Enlightenment science, when he placed science and Scripture in a relation of equivalence by describing them respectively as "the book of God's works" and "the book of God's words," warned that one must not "unwisely mingle or confound these learnings together" lest there arise "not only a fantastic philosophy but also an heretical religion."[11]

Interpreting Scripture Irrespective of Science

For the reasons just laid out, our interpretive method in the chapters that follow will be one that assumes the separateness of science and Scripture, not their complementarity. This means that, in interpreting biblical material relevant to the question of the origin of human physical mortality, we will take no account of modern scientific answers and we are fully open to the possibility of dissonance on this issue between biblical statements and scientific consensus.

As is clear from our discussion so far, not all interpreters approach Scripture in this way. Lucas, for example, recognizing that the sciences of language and history are "indispensable tools" for interpretation, goes on to announce that he "cannot see that there is any difference in principle" with regard to "using the findings of the natural sciences."[12] But there is a large difference in principle: philological study of Scripture (employing linguistics and history) focuses on the *medium* of expression, perhaps thereby exposing the message, whereas natural science can say nothing about the medium, only the *message*—and that by an anachronistic comparison which issues in an unhelpful binary judgement of "true" or "false."

11. Quoted in Hyers, *Meaning*, 32–33.
12. Lucas, "Issues," 46.

Where natural science can play a minimal role in interpretation is in its alerting us to the possibility that certain biblical expressions are figurative or phenomenalist. "It is only acceptance of the scientific evidence for a heliocentric view of the solar system that leads us to say that this [Psalms 19:6, 93:1] is the language of appearance"[13]—and that allows us to disengage science and Scripture on this point. But even here it is not science that directly performs this role, but rather common sense, sometimes informed by scientific discovery.

Accepting Instances of Real Contradiction between the Bible and Science

Being open to the possibility of dissonance between the Bible and science on the issue of human physical mortality (among other matters of natural fact) presents us with the question of what we should do with dissonance if we discover it. We have already discounted the literalist and concordist responses, although on any point of tension between the Bible and science we do well to exercise some of the literalist's scepticism of science (indeed, scepticism is staple to scientific inquiry) and to consider, not unlike the concordist, whether a legitimate harmonization might be possible. If the dissonance persists, two other responses are possible.

The first is to move beyond our methodological assumption of the separateness of science and Scripture, to assert an *ontological* separation between the two. This approach sees the Bible and science as each concerned with distinct realms of being and different levels of description. It holds the biblical authors to have been preoccupied with "primary causes" (ultimate origins and purposes) and science to be taken up with "secondary causes" (interactions within nature); it supposes the one to answer questions of "why?" and the other of "how?"[14] Hyers champions such separatism with reference to the early chapters of Genesis, claiming that they "are

13. Ibid., 48.
14. McGrath, *Theology*, 275.

not in conflict with scientific and historical knowledge . . . precisely because they have little to do with it. They belong to a different literary genre, type of knowledge, and kind of concern."[15]

Although Hyers cogently presents the case that comparing the Bible and science is like comparing, not apples and oranges, but "oranges and orangutans,"[16] and although he offers a plausible alternative for the genre and concerns of Genesis 1–3, we note that he can only go as far as to say that Genesis and science have *little* to do with each other, not nothing at all. For Scripture and science cannot be disengaged completely. Even though the biblical authors gave no explanations, indeed no descriptions, of natural phenomena with scientific precision, we cannot rule out a priori the possibility that they sought, on occasion, to explain the world around them; that they did venture to answer "how?" as well as "why?"[17] Even Hyers concedes this, and thus admits the likelihood of substantive contradiction between the Bible and science: "Insofar as an explanatory element is present in the stories, such as an account of the origin of languages in a divine judgement on building the tower of Babel, it may be said to be supplanted by later [e.g., scientific] understandings."[18] With regard to our topic of interest, a discernible etiology of human physical mortality in any of the relevant biblical literature would potentially set it at odds with the scientific account.

Thus the second possible response to persistent dissonance—and the one that we favor, should the need arise—is to attribute it to real error on the part of the biblical author, though remembering that we may also err in our interpretations of the text or the material evidence, or both, and remembering that, anyway, it may not have been the author's intention to pronounce on a matter of natural fact.

We reject inerrantist handlings of the Bible's depictions of the natural world, on the grounds of a false assumption about use of

15. Hyers, *Meaning*, 28.
16. Ibid., 31.
17. Duce, "Complementarity," 149–50.
18. Hyers, *Meaning*, 105.

language and incorrect attitudes towards science. Exegesis of the biblical literature should be conducted with cognizance of the ancient contexts in which it was produced and without reference to the alien context and conclusions of modern science. However, to the extent that our own questions about nature, such as about the origin of human physical mortality, are questions that the biblical authors also presumed to answer, the Bible may both contradict science and be in error.

When the exegetical task is accomplished, carefully and with candor, the larger theological exercise can begin. Briefly, this involves setting the biblical meanings alongside each other to assess the degree of biblical diversity on the issue in hand, setting them alongside the scientific perspective to discern the level of genuine concordance, and setting them in relation with relevant areas of theological understanding. The goal—not easily achieved alone and even less easily achieved with consensus—is to articulate a singular reality on the aspect in question that reflects a complementarity between science and theology but not necessarily Scripture. Regarding the aspect of reality that is human physical mortality, we cannot expect to reach this distant goal within the confines of this book. Rather, our more moderate aim is to take a fresh look at the relevant biblical material, to divest it if need be of meanings that have been brought to it and that it cannot bear, and to bring the exegetical outcome within the hearing of that other voice on the matter, science.

4

Biblical Perspectives on Death

TRADITIONAL CHRISTIAN ASSERTIONS REGARDING the origin of human physical mortality rest on interpretations of a handful of biblical texts. We will examine two of those texts in detail in forthcoming chapters. However, they are best seen against a broad background of biblical perceptions of death. Any biblical statements concerning the origin of death should be interpreted in the light of what death was understood to be. Surveying this background is no simple task, for two reasons.

First, the subject of death lies at the periphery of Israelite and New Testament faith. Since the biblical focus is on experience of God and exercise of faith during *this* life, reference to death in the Bible takes the form of cultural attitude rather than explicit theological commitment. Second, because the biblical material was composed over a wide chronological and cultural span, there is diversity in its depictions of death. For an issue as complex as death, and for a period as long as that represented by the biblical literature, we should not be surprised to find that "contradiction, not gradual development of beliefs, is the main fact."[1]

---

1. Martin-Achard, *Death*, 16.

Death in the Old Testament

The verb *muwth* ("to die") appears over 600 times and the noun *maveth* ("death") 158 times in the Old Testament.[2] Most of these are plain notations of the form "*N* died" or "the death of *N*" (e.g., 1 Chronicles 1:44–51; 2 Kings 14:17). Although it would seem that these deaths were recorded more for their familial and community significance than out of abstract theological reflection, how these deaths were perceived is of some interest in our concern with OT beliefs about physical death. In addition to the literal sense, there is a rich metaphorical usage of "die"/"death" in the OT: for example, (i) for circumstances that are somehow in opposition to life, such as poverty (e.g., 1 Samuel 2:6–8); (ii) for a state of disobedience against God (e.g., Deuteronomy 30:11–20); and (iii) occasionally as a personified power (e.g., Job 18:13–14).

Generally the attitude implicit in the OT is one of sober acceptance of the inevitability of physical death. Although there is mild complaint about life's brevity (Psalms 39:5, 89:47), there is no protest in the OT against the fact of physical mortality. Unlike suffering, mortality seems not to have raised the question of theodicy.[3] Joshua and David dispassionately anticipate their being "about to go the way of all the earth" (Joshua 23:14; 1 Kings 2:2) and the author of Ecclesiastes writes in matter-of-fact terms of "a time to be born, and a time to die" (Ecclesiastes 3:2).[4] For Job, "a mortal . . . , few of days and full of trouble" (Job 14:1), death is a welcome prospect, bringing relief (Job 3:11–22 cf. Jeremiah 20:14–18). Despite a tradition of two individuals having not died (Genesis 5:24; 2 Kings 2:11), death was perceived in ancient Israel as the unavoidable occasion when the body returned to the earthy source from which it derives (Psalm 103:14; Ecclesiastes 3:20) and when its life-force (*nephesh, ruwach*) ceases to animate it (Genesis

---

2. Bailey, "Death," 23.

3. Bailey, *Perspectives*, 52.

4. Qohelet's protest that "this is an evil" (9:3) is not against mortality itself but against the perceived injustice that "the same fate comes to everyone" regardless of the moral character of their lives.

35:18; Psalm 146:4), returning to God (Job 34:14–15; Ecclesiastes 12:7). Thereafter the dead individual was believed to continue a very minimal existence as a *rephaim* ("shade") under the earth (Psalm 88:5, 10–12; Isaiah 26:14).

In the Israelite conception, inhabitants of the netherworld were thought to be separated from Yahweh: "In the world of the dead, Yahweh's work, Yahweh's proclamation and Yahweh's praise no longer have any place"[5] (e.g., Psalm 115:17; Isaiah 38:11, 18). In itself, this "theological vacuum"[6] was not perceived as a problem. One possible reason for this, and for the general acceptance of mortality, was the value the Israelites placed on community; "since God's relationship is not primarily with individuals but with the people, death does not represent a threat to faith."[7] Nevertheless, the vacuum that was the realm of the dead *was* filled by the extension of another possible "coping mechanism,"[8] a robust sense of Yahweh's sovereignty. The OT contains affirmations of Yahweh's sovereignty over death and the realm of the dead as much as over life and the living (e.g., Job 19:25–26; Psalm 139:8). This "certainty of God"[9] over death became further pronounced in early apocalyptic thinking (see below).

Despite the general acceptance of death—usually when the deceased was "full of years" (Genesis 25:8 [Abraham], 35:29 [Isaac]; 1 Chronicles 29:28 [David]; Job 42:17 [Job])—some ways of dying were decidedly unacceptable: prematurely (e.g., Isaiah 38:1–3, 10, 12), violently (e.g., Ezekiel 28:8–10), or without (male) offspring (e.g., 2 Samuel 14:7). Premature death was often seen as a punishment for sin (e.g., Exodus 32:27, 35; Numbers 16:26–35), not least in those cases where an individual committed an offense to which the death penalty was attached in the Mosaic legislation (e.g., Leviticus 20:1–27). This penal conception was, however, confined to the immediacy of death following on the specific offenses of particular

5. Wolff, *Anthropology*, 106.
6. Ibid., 110.
7. Schmithals, "Death," 433.
8. Bailey, *Perspectives*, 57.
9. Wolff, *Anthropology*, 110.

individuals or groups;[10] except for a single oblique reference (Psalm 90:7–10), no connection was made between general sinfulness and human mortality until the intertestamental period.

In the later OT literature, two incongruous perspectives exist side by side. On the one hand, human physical death will remain a feature of an envisioned future Davidic realm (Isaiah 65:20); there is continued acquiescence in mortal existence. On the other hand, there are signs of an increasing certainty that God will "overcome" death. Several factors may have been behind this trend: first (as mentioned above), "an absolute confidence in Israel's God as all-powerful" over even death;[11] second, an increased individualization of Israelite faith at the expense of a corporate theological identity, rendering death problematic;[12] third, a longing for national restoration on an apocalyptic scale in the face of historical realities after the exile.[13] Three verses are of principal relevance: Isaiah 25:7 ("[Yahweh] will swallow up death forever"); Isaiah 26:19 ("Your dead shall live"); and Daniel 12:2 ("Many of those who sleep in the dust of the earth shall awake"). The extent to which this language is literal or metaphorical, and of personal or national application, is disputed,[14] but there is a consensus that one or other of these verses contains an implicit expectation of the undoing of human physical death, specifically through personal resurrection.

Death in the Intertestamental Literature

An expectation of bodily resurrection, incipient in the latest OT material, became heightened in the intertestamental period, taking "vivid hold in Jewish belief and writing"[15] in the three centuries before Christ. Part catalyst for this was the experience of

10. Johnston, *Shades*, 43.

11. Ibid., 219.

12. Silberman, "Death," 26–30.

13. Bailey, *Perspectives*, 67–74.

14. See discussions in Johnston, *Shades*, 226; Bailey, *Perspectives*, 73–74; Boer, *Defeat*, 44–50; Wright, *Resurrection*, 108–17.

15. Johnston, *Shades*, 229.

persecution under Antiochus Epiphanes (ca. 215–164 BC) which spawned a hope of vindication for the martyrs (see 2 Maccabees 7:9, 11, 14, 23, 29; 12:44–45; 14:46).[16] The oppression also gave rise to an apocalyptic eschatology that envisioned an imminent turn of the ages, to be accompanied by a doing away with death (e.g., 1 Enoch 51:1–2).

It is much more difficult to tease apart the meanings of "death" in intertestamental Jewish apocalyptic eschatology than in the OT, except to discern that, in addition to the literal and metaphorical senses in the OT, another appears: eschatological death as exclusion from the new age.[17] Regarding the origin of the composite "death" that characterizes the present age, de Boer identifies two "tracks" of Jewish apocalyptic eschatology: "cosmological" and "forensic."[18] The first of these implicates an *angelic* fall as the source of present human sinfulness and death (especially of the righteous by bloodshed) (e.g., 1 Enoch 69:4–11). The second track accords these evils to a culpable *human* rejection of God initiated by Adam and/or Eve but perpetuated in the "evil heart" of every person (4 Ezra 3:7, 21–22).[19] Notwithstanding the fact that in both tracks the term "death" connotes a phenomenon that cannot be reduced to physical mortality, it is more likely than not that this literature evidences a belief that human physical mortality originated with sin and continues to stem from a general state of sinfulness.[20] The author of (the intertestamental) Wisdom of Solomon expressed this succinctly: "God created us for incorruption, . . . through the devil's envy death entered the world" (2:23–24).

---

16. This was the background to Daniel 12:2.

17. de Boer, *The Defeat of Death*, 84.

18. Ibid., 85–86.

19. See also: 4 Ezra 7:118–19; 2 Baruch 23:4; 54:15, 19; Ecclesiasticus 25:24.

20. This is most evident in 2 Baruch, in which "Adam's transgression affects the very structure of created being, perverting it both morally and ontologically" (de Boer, *Defeat of Death*, 81). See 2 Baruch 17:2–3; 48:42–43; 56:6.

Death in the New Testament

The diversity of senses and settings in which *thanatos* ("death") is mentioned in the New Testament is as great as in the OT and intertestamental literature put together. We cannot do justice to this variety here; we are limited to a thumbnail sketch of NT perspectives most pertinent to our interest in human physical mortality.

First, in continuity with the prevailing OT view, we notice that in many instances physical mortality is accepted without demur by those who populate the Gospels (Matthew 8:22; 22:23–28; Luke 2:29), though the loss of a loved one was sorely felt and mourned in the customary manner (Mark 5:38). In the Synoptic Gospels, human physical death is for Jesus never a problem to be resolved (see Matthew 8:22), though on occasion he temporarily reversed it (Matthew 9:18–19, 23–24; Luke 7:1–10). The uncertain length of life served to underline the urgency of his message (Luke 13:1–5). Though his audience may have thought calamitous death to be divine punishment for sin, Jesus emphatically did not (Luke 13:1–5; Matthew 13:24–30). Physical death was but the prelude to either an ultimate terrible eventuality (Matthew 10:28 cf. 5:29–30) or a glorious future reward (Luke 16:9 cf. 14:13–14). These references to eschatological death and life lie within the framework of intertestamental apocalyptic, represented most clearly in the NT by the envisioning in Revelation of "the second death" (Revelation 2:11; 20:13–14) and a time when "death will be no more" (21:4).

The Fourth Gospel is distinguishable from the Synoptics in that, in addition to neutral reference to physical death (John 11:13; 12:33; 18:32; 21:19), "death" and "die" sometimes have a metaphorical sense. This sense is set alongside the literal meaning in Jesus' seemingly paradoxical statement that believers, "even though they die, will live; and everyone who lives and believes . . . will never die" (John 11:25–26 cf. 8:51–53). "[Eternal] life," with both a present and a future aspect, is the "central soteriological notion"[21] in the Fourth Gospel. Even when not apposed to "death"/"die," the implication is that "death," being the opposite

21. Thomas, "Meaning," 204.

of "[eternal] life," is "a quality of existence which the followers of Jesus are able to transcend."[22]

Paul acknowledges human physical mortality, including his own, numerous times in his letters (e.g., Romans 1:23; 2 Corinthians 4:11),[23] but he also expresses a more theological perspective on "death" than is found in the Synoptics or even in the Fourth Gospel. "Death" was for Paul profoundly unacceptable: a corollary of sin and an agent "reigning" despite God (Romans 5:12–21; 1 Corinthians 15:26). This perspective is shared by the authors of several other NT epistles (Hebrews 2:14–15; James 1:15; 1 John 5:16b). Its depiction as a paradigm of present existence distinguishes this "death" from the future perdition of the Synoptics and Revelation, but the extent to which it correlates with physical mortality, if at all, demands careful inquiry.[24]

The difficulty of establishing what "death" connotes or denotes in any particular instance stems in large part from the NT focus on the multiplicity of significances of one physical death and rising from death—that of Jesus. For example, the association of "death" with personified evil may stem from the fact that Jesus was *put to* death (Romans 6:9 cf. 1 Corinthians 2:8), and its association with sin has to do with the understanding of Jesus' death as an *atoning* death (Romans 3:25). Thus an array of metaphorical meanings for "death" attends the NT theological interpretations of Jesus' death—at base a physical death—raising the question of their relation to the physical death that Christians still die. Answering this question requires an awareness that the senses of "death" in the NT are not only diverse but also interconnected and overlapping, and calls for an openness to ambiguity for any particular NT mention of death.

---

22. Bailey, *Perspectives*, 94.

23. See also Romans 4:19; 6:12; 7:2–3; 14:8; 2 Corinthians 5:4.

24. See chapter 6.

5

Genesis 2–3 and Human Physical Mortality

THE FIRST MENTION OF dying in the Bible occurs as early as in its second and third chapters (Genesis 2:17, 3:3–4). Since dying is mentioned in proximity to the origin of "the heavens and the earth" (1:1) and of the first human being (2:4b–7), we could expect to find here an implication, at least, on the origin of death. It is wise to be cautious about bringing such an expectation to the text, however; even if Genesis 2–3 is about origins, it is unlikely to be about the origin of everything it mentions. Nevertheless, if only because of the prominent place these chapters have had in both doctrinal pronouncement and popular discussion on the origin of death, they warrant close study in this connection.

The question of what Genesis 2–3 is "about" has dominated Christian reflection on these chapters. They have been read primarily out of interest in the external realities to which the text may refer. This interest, coupled with our perennial curiosity in origins, especially our own, has caused Genesis 2–3 to be mined for all that it purportedly says "about" beginnings, whether the beginning of physical entities such as the human body or of theological dispositions such as sin. Regrettably, there has often been insufficient recognition of the literary form of Genesis 2–3, which is the form of a story. Whatever else a story may be "about," it is most obviously about the characters that populate it: their actions, reactions, attitudes, and interrelations. Stories require that we look first for meaning *within* them rather than behind them or in front of them; they are "portraits with a world of their own" before they are windows or mirrors.[1]

Our initial task in interpreting the narrative that is Genesis 2–3, then, is to enter its world. Worlds, even literary ones, are large places, and we are interested in just one feature: physical death. We must beware lest our singular interest distort our perception of the story world as a whole and the place of death in it. Being large, narrative worlds can never be characterized in full; there are always gaps in their depiction. The reader must fill at least some of the gaps with information from outside the story. Insofar as ambiguity surrounds the portrayal of death within Genesis 2–3, we will need to consider factors external to the narrative that may bring greater clarity to that portrayal: issues of composition, such as the historical setting in which the text arose, or literary sources the author may have used. With that, we will have stepped outside the world of the story and begun to relate it to the world we inhabit.

Genesis 2:4—3:24

[4] These are the generations of the heavens and the earth when they were created.

Another Account of the Creation

In the day that the Lord[a] God made the earth and the heavens, [5] when no plant of the field was yet in the earth and no herb of the field had yet sprung up—for the Lord God had not caused it to rain upon the earth, and there was no one to till the ground; [6] but a stream would rise from the earth, and water the whole face of the ground— [7] then the Lord God formed man from the dust of the ground,[b] and breathed into his nostrils the breath of life; and the man became a living being. [8] And

---

1. Goldingay, *Interpretation*, 5–6.

the Lord God planted a garden in Eden, in the east; and there he put the man whom he had formed. ⁹ Out of the ground the Lord God made to grow every tree that is pleasant to the sight and good for food, the tree of life also in the midst of the garden, and the tree of the knowledge of good and evil.

¹⁰ A river flows out of Eden to water the garden, and from there it divides and becomes four branches. ¹¹ The name of the first is Pishon; it is the one that flows around the whole land of Havilah, where there is gold; ¹² and the gold of that land is good; bdellium and onyx stone are there. ¹³ The name of the second river is Gihon; it is the one that flows around the whole land of Cush. ¹⁴ The name of the third river is Tigris, which flows east of Assyria. And the fourth river is the Euphrates.

¹⁵ The Lord God took the man and put him in the garden of Eden to till it and keep it. ¹⁶ And the Lord God commanded the man, "You may freely eat of every tree of the garden; ¹⁷ but of the tree of the knowledge of good and evil you shall not eat, for in the day that you eat of it you shall die."

¹⁸ Then the Lord God said, "It is not good that the man should be alone; I will make him a helper as his partner." ¹⁹ So out of the ground the Lord God formed every animal of the field and every bird of the air, and brought them to the man to see what he would call them; and whatever the man called every living creature, that was its name. ²⁰ The man gave names to all cattle, and to the birds of the air, and to every animal of the field; but for the man[c] there was not found a helper as his partner. ²¹ So the Lord God caused a deep sleep to fall upon the man, and he slept; then he took one of his ribs and closed up its place with flesh. ²² And the rib that the Lord God had taken from the man he made into a woman and brought her to the man. ²³ Then the man said,

"This at last is bone of my bones
and flesh of my flesh;
this one shall be called Woman,[d]
for out of Man[e] this one was taken."

²⁴ Therefore a man leaves his father and his mother and clings to his wife, and they become one flesh. ²⁵ And the man and his wife were both naked, and were not ashamed.

The First Sin and Its Punishment

3 Now the serpent was more crafty than any other wild animal that the Lord God had made. He said to the woman, "Did God say, 'You shall not eat from any tree in the garden'?" ² The woman said to the serpent, "We may eat of the fruit of the trees in the garden; ³ but God said, 'You shall not eat of the fruit of the tree that is in the middle of the garden, nor shall you touch it, or you shall die.'" ⁴ But the serpent said to the woman, "You will not die; ⁵ for God knows that when you eat of it your eyes will be opened, and you will be like God,[f] knowing good and evil." ⁶ So when the woman saw that the tree was good for food, and that it was a delight to the eyes, and that the tree was to be desired to make one wise, she took of its fruit and ate; and she also gave some to her husband, who was with her, and he ate. ⁷ Then the eyes of both were opened, and they knew that they were naked; and they sewed fig leaves together and made loincloths for themselves.

⁸ They heard the sound of the Lord God walking in the garden at the time of the evening breeze, and the man and his wife hid themselves from the presence of the Lord God among the trees of the garden. ⁹ But the Lord God called to the man, and said to him, "Where are you?" ¹⁰ He said, "I heard the sound of you in the garden, and I was afraid, because I was naked; and I hid myself." ¹¹ He said, "Who told you that you were naked? Have you eaten from the tree of which I commanded you not to eat?" ¹² The man said, "The woman whom you gave to be with me, she gave me fruit from the tree, and I ate." ¹³ Then the Lord God said to the woman, "What is this that you have done?" The woman said, "The serpent tricked me, and I ate." ¹⁴ The Lord God said to the serpent,

"Because you have done this,
cursed are you among all animals

and among all wild creatures;
upon your belly you shall go,
and dust you shall eat
all the days of your life.
¹⁵ I will put enmity between you and the woman,
and between your offspring and hers;
he will strike your head,
and you will strike his heel."

¹⁶ To the woman he said,

"I will greatly increase your pangs in childbearing;
in pain you shall bring forth children,
yet your desire shall be for your husband,
and he shall rule over you."

¹⁷ And to the man[g] he said,

"Because you have listened to the voice of your wife,
and have eaten of the tree
about which I commanded you,
'You shall not eat of it,'
cursed is the ground because of you;
in toil you shall eat of it all the days of your life;
¹⁸ thorns and thistles it shall bring forth for you;
and you shall eat the plants of the field.
¹⁹ By the sweat of your face
you shall eat bread
until you return to the ground,
for out of it you were taken;
you are dust,
and to dust you shall return."

²⁰ The man named his wife Eve,[h] because she was the mother of all living. ²¹ And the Lord God made garments of skins for the man[i] and for his wife, and clothed them.
²² Then the Lord God said, "See, the man has become like one of us, knowing good and evil; and now, he might reach out his hand and take also from the tree of life, and eat, and live forever."— ²³ therefore the Lord God sent him forth from the garden of Eden, to till the ground from which he was taken. ²⁴ He drove out the man; and

at the east of the garden of Eden he placed the cherubim, and a sword flaming and turning to guard the way to the tree of life.

". . . in the Day You Eat of [the Tree], You Shall Die."

The story in Genesis 2:4b—3:24 begins and ends outside a garden called Eden. Nevertheless, we pick up the story within the garden, with the first mention of dying in 2:17: [Yahweh God commanded,] "of the tree of the knowledge of good and evil you shall not eat, for in the day that you eat of it you shall die."

This verse consists of two clauses: the first, [not + infinitive] ("you shall not eat"), as also found in the Ten Commandments (Exodus 20:3–17); and the second, a motive clause [infinitive + qal] ("you shall die"), as found in threats in OT historical narrative and prophetic texts (e.g., Genesis 20:7; 1 Samuel 14:39, 44).[2] God's utterance in this verse is therefore properly understood as a prohibition and a penalty, not a piece of advice against eating lethally poisonous fruit.[3] It is also instructive to note that the penalty pronounced here is one of dying, not of becoming mortal; there is no adjective or noun in biblical Hebrew corresponding to "mortal(ity)."[4] The final linguistic feature of 2:17 that needs our attention is the time reference attached to the penalty: "in the day [that you eat of it]." Modern commentators agree that this is a Hebrew idiom, chronologically indefinite, with a meaning such as "when" (as in 2:4; 5:1)[5] or "promptly" (see Numbers 30:6, 8, 9).[6]

Since the narrative stresses the commonality between the man and the woman (2:24), it is safe to assume that the prohibition, although delivered to the man, applied equally to both of them.

2. Wenham, *Genesis*, 67. Note, however, that [infinitive + *hophal*] is used in legal texts pronouncing the death penalty (e.g., Leviticus 20).

3. So Beattie, "Genesis 2–3," 8–10.

4. Moberly, "Serpent," 15–16.

5. Ibid., 14.

6. Wenham, *Genesis*, 68.

And both broke it: "She took . . . and ate . . . and he ate" (3:6). But whereas the eating clearly recalls the prohibition, the first result of eating, the "opening of their eyes" (3:7), bears no resemblance to the penalty of dying that God had pronounced. The couple live on in the verses that follow. Clearly, the serpent was correct in his prediction that the couple's eyes would be opened (3:5), and initially it would appear that he was also correct in disputing God's penalty of dying (3:4). Moberly makes a fair assessment when he suggests that "the crux for the interpreter must surely be the resolution of this remarkable anomaly."[7]

The anomaly consists in an unthinkable eventuality: that, by the midpoint of the story, God, who in the initial sequence supremely creates all (2:4b, 5, 7–9, 19, 22), orchestrates all (2:8, 15, 18, 19b–21), and officiates over all (2:16–17), is seemingly trumped by one of God's mere creatures, albeit the most crafty. But the story itself is not anomalous; stories are not stories without such twists, which are not just twists *of* the plot but generate the need *for* a plot.[8] The anomaly is one of several others in the story, and the narrative resolves those others itself. The man's unsatisfactory solitude is met with "flesh of [his] flesh" (2:18–24) and the couple's discovered nakedness is overcome by "garments of skins" (3:7, 21). The first place to look for a resolution of the somewhat larger anomaly of an apparently abortive divine penalty is therefore the remainder of the narrative.

Some conceivable resolutions can be ruled out: we read nothing of God modifying the penalty[9] or of the couple reforming themselves. Besides, these outcomes would amount not just to abortion of the penalty but to an abortion of plot—unless the story is a literary tragedy, recounting the failure of God and not, as traditionally understood, the "fall" of the couple. The avenue open to us is that "God's death sentence was carried out, but in some

---

7. Moberly, "Serpent," 9.

8. Wright, *New Testament*, 71.

9. So Skinner, *Genesis*, 67.

way other than the obvious and straightforward way that his words initially implied."[10]

". . . until You Return to the Ground"

The curses and afflictions that God pronounces in 3:14–19 are the first candidates to appear in the story for a penalty corresponding to the one foretold in 2:17. A connection with the prohibition, even if not the penalty, is indicated by the restatement of it in 3:17a. But, since the words "die" or "death" are nowhere used, we are forced to look closely for the "some other way" of correspondence between the penalty originally announced and the one perhaps here applied.

Of the afflictions that God pronounces separately on the woman (3:16) and the man (3:17–19), the final clauses of God's word to the man draw our attention most of all: he will "return to the ground . . . to dust." (3:19b). This is the clearest allusion to dying since the couple's offense. Eager to locate the imposition of the penalty of dying, some commentators understand 3:19b as applying to the woman as well as to the man. Blocher, for example, contends that "since [the woman] shares the man's humanity, she too will experience the penalties suffered by mankind in general."[11] But none of the penalties is addressed to "mankind in general," and we may imagine that any suffering shared by the man and the woman may be due not to their "sharing humanity" but rather to their being in close relationship, such that neither could be unaffected by each other's sex-specific affliction. To understand "until you return to the ground, for out of it you were taken" as also applying to the woman would require the woman to have also been originally taken from the ground, contrary to the story. It is only indirectly, by way of derivation from the man (2:21–23), that the woman, like the man, derives from the ground. The woman was

---

10. Moberly, "Serpent," 13.

11. Blocher, *Beginning*, 180. See also Westermann, *Genesis 1–11*, 263.

neither taken from the ground nor put to till it; she is subject to the man's afflictions no more than he is to hers of pain in childbearing.

Notwithstanding the sex specificity of "until you return to the ground," whether we have here the enactment, at last, of God's original penalty of dying or merely a statement of the man's original physical impermanence, it is not possible to tell, at least on the evidence internal to the narrative. On the one hand, (i) the repetition of the prohibition (3:17) brings to mind the penalty of "you shall die," conceivably in preparation for its endorsement in the subsequent verses,[12] and (ii) the absence of mention of "breath of life" speaks against a rhetorical parallel with 2:7.[13] On the other hand, (i) there is no explicit reference to the death penalty of 2:17, and (ii) the mention of both "ground" and "dust" is very probably a "reminiscence of the man's creation."[14]

Johnston would have us believe that the matter can be settled by logical deduction from the concluding sentence of 3:19, "you are dust, and to dust you shall return":

> If this is a simple statement of fact, its conclusion ("to dust you shall return") asserts man's intrinsic mortality. And if it is a statement of divine judgement, its explicit premise ("you are dust") implies human impermanence. So whether read as a statement [of fact] or as a judgement, God's pronouncement in 2:17 implies that man is naturally mortal.[15]

But this amounts to a sleight of hand. With regard to the first part of Johnston's claim, the man's being dust as a matter of fact does not make his eventual dissolution into dust inevitable. Regarding the second part of his claim, the "premise" implies nothing about impermanence or permanence, only that the judgement takes a form appropriate to the man's physical composition. Furthermore, the story thus far does not permit talk of "mortality" (still less the mortality of "man"), for it raises in the reader's mind only the

12. Wenham, *Genesis*, 83.

13. Ibid., 83.

14. Wolff, *Anthropology*, 115.

15. Johnston, *Shades*, 40–41.

question of how and when the couple will experience God's penalty of death, not of whether there is a limitation on their lifespan, either intrinsic or imposed.

"You Shall Die," or "You Shall Suffer until You Die"?

Since that part of God's judgement on the couple which most closely matches the penalty of dying (3:19b) cannot be ruled in or out as a substantive part of the judgement, we must continue our search for God's application of his penalty. Perhaps, though, the penalty has already been announced yet has escaped our notice. Moberly surmises that the couple succumbed to a "qualitative death"[16] almost before they had swallowed their mouthfuls, with alienation from God (3:8–10) and each other (3:12) and, later, in the "public dying" of painful childbirth and painful toil (3:16–19).[17]

It is certainly conceivable that such a metaphorical death-in-life has a place under the heading of the penalty of 2:17; after all, it is not easy to see a relevance for 3:8–19a if these verses are not part of the sentence. It is also conceivable that the "returning to dust" of 3:19b is of a piece with the hardship of 3:16–19a, such that "[literal] death is not a punishment but the term of [the man's] toilsome work,"[18] although Young's related claim that "the ground will at last overcome him [the man]"[19] can be dismissed because the only enmity mentioned is between the offspring of the snake and of the woman, not between the man and the ground. However, this metaphorical application of the penalty leaves us looking for others. God's clear statement "you shall die" (2:17) so soon after giving the "breath of life" (2:7) left us expecting more than the couple's slow decline. But the narrator has not finished yet; there are four more verses to come.

16. Moberly, "Serpent," 16.
17. Ibid., 17.
18. Westermann, *Genesis 1–11*, 267.
19. E. J. Young, quoted in Blocher, *Beginning*, 183.

". . . He Might . . . Take also from the Tree of Life"

It is no surprise that the story continues beyond 3:19, for the curses of 3:14–19 are there merely announced, not actualized. Furthermore, the contrast between the impoverishment of 3:14–19 and the abundance of Eden (2:9–10), and the fact that 3:19 more than likely recalls the man's having been taken from ground *outside* Eden, lead one to expect that actualization of the curses will not be within the garden.

The couple's expulsion from Eden prevents their access to the tree of life (3:22–24). In connection with death, there are two aspects of this tree to consider: the couple's access to it until they were barred, and the likely effect of eating its fruit. The most satisfactory position on the first of these is that the couple, despite having been free to eat from the tree of life, did not. This is the implication of 3:22; since a single sampling of the fruit of the tree of the knowledge of good and evil had resulted in the man "knowing good and evil" (3:22a), we can infer that the fruit of the tree of life would have conferred "[life] forever" (3:22b) on first bite, and, this being disallowed, no bite was taken. Blocher ignores this logic of 3:22, instead mistaking a permission to eat from any tree except one (2:16) for a command that was obeyed.[20]

As for the second aspect (the likely effect of the tree's fruit), having asserted that 3:22 refers to a one-off eating of the tree of life, we have rejected one imaginable effect: nutritional sustenance, such that the couple, being barred from the tree, lost a lifeline. This is not to say, however, that the tree of life was of no relevance to their physical constitution; the couple's embodied existence has been underlined multiple times in the narrative so far (e.g., 2:7, 23, 3:19b), so the "living forever" that would result from eating the second tree's fruit would almost certainly be continued physical life, not disembodied "life after death." Here, then, for the first time

20. Blocher, *Beginning*, 122. Grammatically, 2:16 is a command, but the inclusion of an emphatic verbal form ("freely") and the wide application "every [tree]" indicate permission not obligation (Moberly, "Serpent," 3–4).

in the narrative, we encounter the matter of the couple's physical mortality or immortality, beyond the question of their dying or not in punishment. The clear implication of their being barred from the tree is that, at least just prior to that moment, they were physically mortal. The key question for us is whether, had they "reach[ed] out and take[n] also from the tree of life" (3:22), they would have gained a bodily immortality that *they had never had before*, or regained a physical immortality that *they once had but lost* in judgement.

These two logical alternatives correlate with the pair presented earlier for 3:19b, and it is no easier to decide between them. The first alternative would bring our search for the enactment of God's penalty to an end only if a warning as firm as "you shall die" (2:17) could be fulfilled by a punishment as legally fictitious as the mere loss of *potential* immortality. The second requires a back-reading of 2:7; 2:17; and 3:19 in terms of a conception of (im)mortality absent from those verses. Concerning the tree of life, the story leaves the reader so entirely free to speculate on both prospective and retrospective "what ifs" that we are at a loss to derive from the tree's presence any clue as to God's enactment of his penalty and the couple's original state of mortality or immortality.

"[God] Drove Out the Man"

The garden has been the literary setting of the story since the opening sequence, so when the man is banished from it the reader also experiences something of his removal. The man's being sent to "till the ground from which he was taken" (3:23) also takes the reader abruptly out of the somewhat familiar, if enigmatic, garden surroundings to the lesser known and little remembered "ground" before and beyond it. The last verse (3:24), in which the man is emphatically "dr[iven] out" and the tree of life impenetrably guarded, brings a finality to the man's lot and an end to any hopes the reader may have had for him and his companion. Could it be that "the movement of the unit casts the primary punishment for the disobedience . . . as expulsion from the garden"?[21] If so, the original penalty would again transpire to be one of metaphorical, not literal, "dying," for the couple live on, even to beget new human life (3:20; 4:1).

Whether the expulsion itself does better service as the penalty of "dying" than the afflictions already announced (3:16–19) depends on the significance of the garden; "dying" can only be a metaphor for expulsion from a garden if the "garden" is in turn a metaphor for a reality that has to do with life in some sense. To establish this, we too need to cross the garden fence and relate the garden, its contents, and the events that take place within it to the world outside the story, the world in which it was written.

The garden of Eden, with its talking serpent and its unusual trees, first strikes us as more of a fictional than a factual place, but the borderline between these two categories is a blurred one. Fiction "is based on factual human experience and without this would neither exist nor be intelligible or interesting."[22] The human experience closest to the Eden story was that of its first authors and audience. Knowing the historical context in which the story was composed, then, offers the best hope of rendering it intelligible, not least those aspects of it that may allude to death.

The Eden narrative was most probably written at a fairly late stage in Hebrew cultural development.[23] Apart from the mature Hebrew language in which it is written, there are two other indications of this: the analogy with covenant theology, and the sanctuary symbolism. Regarding the first of these, the likeness between (in the story) God's bountiful provision for the couple but his proscription of a particular act, and (in Israelite religion) placement in the Land but the requirement to obey Torah, suggests that Genesis 2–3 is the result of a process of reflection on the latter. Torah obedience or disobedience had been expressed in Deuteronomy in terms of the consequence of either qualitative "life" or

21. Coats, "God," 231.

22. Goldingay, *Scripture*, 73.

23. Moberly, "Serpent," 1–2.

"death" (Deuteronomy 30:15–19),[24] and the Eden narrative could be a literary reworking of that message.[25]

Regarding the second indication of historical context, the sanctuary symbolism consists in: (i) the location of the garden "in the east" (2:8); (ii) a possible correspondence between the trees and items used in Israelite worship; (iii) the mention of precious materials used in priestly vestments (2:12); and (iv) in the final verse of the story (3:24), "a remarkable concentration of powerful symbols [cherubim, fire, 'the east'] that can be interpreted in the light of later sanctuary design."[26] Whether the sanctuary symbolized by Eden was the tabernacle or the temple need not concern us here; both were places where Yahweh was specially present, and to be put out of Yahweh's presence, as the man was from Eden, was to die a "death" no less serious for its being a metaphorical, not a literal, one.[27]

Here, we suggest, in the significance of the couple's banishment from the garden, is the consummation of the penalty of dying that God had warned he would apply. It is a significance that would have been readily apparent to those who wrote and first heard the story, unlike to us, who are, perhaps, more dimly aware of our obligation to obey our maker and of our need for his sustaining presence.

A Message about Matters Other than Physical Death

Based on the symbolism of Genesis 2–3 just outlined, the story can be understood in its historical context as a parable of core elements of Hebrew religion and/or as a prefiguring of what did eventuate: the exile of both northern and southern kingdoms from the Land. In neither of these understandings is the divine penalty

24. See also Isaiah 24:4–6; Jeremiah 21:8; Proverbs 14:12; 16:25.

25. Wenham, *Genesis*, 90; Moberly, "Serpent," 16.

26. Points i–iv are from Wenham, *Genesis*, 86.

27. Ibid., 74. See Leviticus 13:45–46; Numbers 5:2–4; 1 Samuel 15:35; Proverbs 8:35–36.

in Genesis 2–3 one of literal, physical death. This is due partly to the constraints of the storytelling and partly to a theological purpose of the author. As regards the storytelling, the continuity between the first couple and subsequent generations would have been broken had the author allowed the couple to be struck down dead before parenthood rather than expelled to a metaphorical, living "death" with offspring. Without that continuity, Genesis 2–3 could not form part of a preface to the chronicle of the patriarchs that begins in chapter 11.

Regarding his theological purpose, wanting to portray Yahweh as true to his word ("you shall die") yet full of grace, the author mitigates the divine penalty in a pattern of repeated mitigation in his prehistory that is Genesis 2–11. Thus, just as the man's future is spared, albeit now affected by his disobedience, so the murderer Cain receives Yahweh's protective mark (4:15). Likewise, the human race, though universally wicked (6:5), is preserved through one of its members, Noah, only to persist in its evil (8:21) but to receive renewed divine blessing (8:22). Genesis 2–11 "sets forth the salutary theological lesson that only a hair's breadth separates God's justice and his mercy"[28]—and that Israel's experience bore greater testimony to the second of these.

This lesson is one that is easily comprehensible outside ancient Israel. Indeed, the lack of explicit reference to covenant, Torah, or temple in Genesis 2–3 allows the story to be applied to a much wider context than the original one and permits a variety of meanings to be discerned in its richness. The ambivalence of the usage of *adam* for both humankind (e.g., 2:5) and an individual character (e.g., 2:18) also invites a wider application. But whether the story proffers explanations of such universal facts of human existence as the complementarity—yet conflict—between the sexes, our ability to make autonomous moral choices, the price of knowledge, our need for clothing, or the common fear of snakes (or any combination of these and others), the vagueness of its mention of death indicates that the story does not also provide an etiology of human physical mortality.

28. Gibson, *Genesis*, 193.

The closest the narrative gets to an etiology of death is in its clear etiology of hardship. But, while it frames the painful tribulation of bearing new life and the heavy travail of sustaining current life as consequences of disobeying the Creator, the narrative stops short of unambiguously including physical death among those consequences. This is, however, only to be expected from ancient Israel's perception of the problem of theodicy; on the biblical evidence, it was a problem raised only by suffering experienced in life, not by the fact of life's finitude.[29]

Mesopotamian Myths as an Interpretive Aid?

Although, in our judgement just expressed, the lack of articulation of an etiology of death in Genesis 2–3 strongly suggests that this issue lay outside the author's focus, there are indications that it was not out of mind. These indications come to light against the background of ancient Near Eastern mythologies.

Mortality is a principal theme in the *Epic of Gilgamesh*.[30] Gilgamesh, grieved by the death of his friend Enkidu, seeks out the only immortal man, Utnapishtim, to quiz him about the secret of his deathlessness. In answer, Utnapishtim relates how the god Enlil had sent a deluge to smother humankind and so gain respite from the noise people produced. Utnapishtim escaped the destruction, having been tipped off by his personal deity, Ea, beforehand. Despite Enlil's initial chagrin on discovering that his genocidal plan had been thwarted, Enlil conferred immortality on Utnapishtim. In the poem of *Atrahasis* there is an addition to the story at this point: the noise of the still-too-numerous humans (who, like Gilgamesh himself, have oddly survived the flood) continues to bother the gods. The solution this time: Ea initiates a curb on population growth by means of hindrances to human fecundity. Back in the *Epic*, Utnapishtim sends Gilgamesh on his way with dashed hopes that he too may be immortalized—except, that is,

29. See chapter 4 above.

30. The following summary is based on that in George (trans.), *Gilgamesh*, xliii–xlvi.

if he were to retrieve a certain plant that can restore one's youth. Gilgamesh succeeds in finding the plant, only to have it promptly snatched away by a serpent before it takes effect.

There is in all this both perplexity at human mortality and an explanation of its origin. If the writer of Genesis 2–3 was aware of these features of the Mesopotamian story—and the relationship between the biblical and Babylonian flood stories makes it likely that he was—then the Eden narrative may contain a response to them, or at least a reflection of them. Regarding perplexity at mortality, "the granting of immortality to one person in the past, and one person only, serves if anything to underline the mortality of all the rest of them—and Gilgamesh's failure to cash in on Utnapishtim's good fortune merely rubs in the message further."[31] In Genesis 2–3 the same message is conveyed by the couple's being barred from the fruit of the tree of everlasting life, which was once there for the picking.

Regarding an explanation of death's origin, if the Genesis writer had one in mind at all, and if he modeled it on that in *Atrahasis*, then "to dust you shall return" (3:19b) is (i) a divine decree (like that initiated by Ea), not a statement of original human mortality, and (ii) an expedient intervention in human existence (like that needed to secure quietude for the gods), not a punishment meted out. The Eden etiology would differ from its *Atrahasis* predecessor in that the advent of death is occasioned by "the first human sin, not humanity's din."[32]

We could, then, enlist the Mesopotamian background to tip the balance of our reading of Genesis 2–3 towards the story implying an original human immortality, humans becoming mortal only subsequent to sin though not in judgement for it. But we remain circumspect about this way of arriving at an interpretive decision. Given that the textual correspondence between Genesis 2–3 and ancient Near Eastern literature amounts to little more than the possibly coincidental featuring in both of a serpent and a plant, it is of doubtful legitimacy to bring a Mesopotamian etiology of

31. Gibson, *Genesis*, 194.
32. Ibid., 181.

death to bear on the relevant verses of Genesis 2–3. Even if the writer was familiar with the perception of mortality in the Mesopotamian mythology, he chose not to make an etiology of physical death even implicit in his story. The most we can say is that he may have had an origin for human physical mortality in mind, but that he did not express his mind on the subject.

Conclusion

The couple in the story that is set in a garden called Eden clearly break a divine prohibition but do not clearly receive the penalty of dying that, in the story, God had said would follow on their disobedience. The story depends for its dramatic tension on an expectation of the reliability of a forewarning from the mouth of Yahweh God, no less, and on the fulfilment of that expectation. So it compels us to look closely for a penalty that amounts to one of dying. God's pronouncement that the man will "return to the ground . . . to dust" (3:19b) and the prevention of access to the tree of life are obvious candidates for this penalty, but, on reflection, we can be no more than ambivalent about their inclusion (or not) in the sentence. They may, in the light of Mesopotamian mythologies, represent a marginal concern with the origin of death, but it is doubtful whether such literature had any influence on the composition of Genesis 2–3. The other candidates would make for a penalty of metaphorical, not literal, death. Given the strong indications that Eden symbolizes Israel's holy sanctuary of God's presence, the couple's expulsion from Eden (3:23–24), more than their becoming subject to afflictions (3:16–19a), has the gravity of a penalty of dying. The story is taken up with the central importance of humans' relationship with their Creator and Provider, and the consequences for that relationship when humans disrupt it. The properties of the human body, whether now or once upon a time, are not germane to this more pressing concern.

6

The Apostle Paul and an Origin for Physical Death

FOR ALL THE MANY mentions of death in the NT, few relate to its origin. Just two passages stand out (both from Paul), not least for their talk of death having "come through" Adam: Romans 5:12–21 and 1 Corinthians 15:1–57. Although exegesis of 1 Corinthians 15 belongs in any complete handling of our topic, we will bypass that chapter, focusing instead on the passage in Romans. The reason for our choice is that, although physical death is certainly at issue in 1 Corinthians 15, Paul's principal concern there is eschatological, not protological. There is an isolated protological mention of death in verse 21, but this has a close parallel in Romans, where Paul articulates the point more fully.

Romans 5:12–21

Adam and Christ

> [12] Therefore, just as sin came into the world through one man, and death came through sin, and so death spread to all because all have sinned— [13] sin was indeed in the world before the law, but sin is not reckoned when there is no law. [14] Yet death exercised dominion from Adam to Moses, even over those whose sins were not like the

transgression of Adam, who is a type of the one who was to come.

¹⁵ But the free gift is not like the trespass. For if the many died through the one man's trespass, much more surely have the grace of God and the free gift in the grace of the one man, Jesus Christ, abounded for the many. ¹⁶ And the free gift is not like the effect of the one man's sin. For the judgment following one trespass brought condemnation, but the free gift following many trespasses brings justification. ¹⁷ If, because of the one man's trespass, death exercised dominion through that one, much more surely will those who receive the abundance of grace and the free gift of righteousness exercise dominion in life through the one man, Jesus Christ.

¹⁸ Therefore just as one man's trespass led to condemnation for all, so one man's act of righteousness leads to justification and life for all. ¹⁹ For just as by the one man's disobedience the many were made sinners, so by the one man's obedience the many will be made righteous. ²⁰ But law came in, with the result that the trespass multiplied; but where sin increased, grace abounded all the more, ²¹ so that, just as sin exercised dominion in death, so grace might also exercise dominion through justification[a] leading to eternal life through Jesus Christ our Lord.

A Multidimensional Passage

The letter to the Romans contains over half of the ninety-five occurrences of *thanatos* ("death") and its cognates in the undisputed Pauline corpus, and the majority (forty-two) of these are in chapters 5–8.[1] Of principal interest in our pursuit of Paul's perspective on the origin of human death is Romans 5:12, in which he remarks that death "came into the world" and "spread to all." As only the first half of an interrupted sentence,[2] this verse belongs within a tightly

1. Black, "Perspectives," 413. There are just seven occurrences outside these chapters: 1:32; 14:7, 8 (three times), 9, 15.

2. Most commentators think that Paul provides the apodosis for the protasis of verse 12 in verse 18, but this is disputed.

written paragraph of ten verses (vv. 12–21). By dint of its prominent Adam–Christ antithesis, this whole paragraph "depends for
its force on Paul's underlying theology of what human beings are
in the divine intention and purpose,"[3] and so demands attention in
any discussion of Paul's conception of human mortality.

Our particular interest is in whether Paul thought *physical*
mortality to be the original divine intention for human beings. It
is therefore important that we establish Paul's meaning for "death"
in Romans 5:12–21. Some commentators waste no time in drawing conclusions as to the kind of death that Paul says "came" and
"spread": that it was indeed physical death,[4] or only spiritual,[5]
or both,[6] or of a kind that is best expressed in other terms.[7] We,
however, resist an inclination to decide on the sense in which Paul
used "death" in this passage until our discussion of it is well ad-
vanced. The complexity and subtlety of Paul's argument warns us
against setting off too quickly on the trail of a singular quest for the
meaning of "death" in 5:12–21.

In fact, Romans 5:12–21 is less a straight-line argument than
an on-the-spot theological pirouette, turning an idea full circle
and, in doing so, showing off its several facets. "Just as . . . so also":
altogether, Paul brings six dimensions of his Adam–Christ typol-
ogy rapidly into and out of view. In keeping with this multidimen-
sional form of the passage, we will examine three aspects of Paul's
portrayal of "death" here: (i) its relation with sin; (ii) the personification of "death"; and (iii) its apposition to "life." By building up
a picture of Paul's conception in this way, we will put ourselves in
a position to judge what place, if any, physical death had in his
conception.

---

3. Wright, *Resurrection*, 249.

4. Ziesler, *Romans*, 145; Morris, *Epistle*, 229.

5. Berry, "Earth," 34.

6. Dunn, *Romans*, 273; Fitzmyer, *Romans*, 412.

7. Barth, *Epistle*, 166: "the supreme law of the world in which we live . . . the
supreme tribulation in which we stand."

"Adam" in Romans 5

Before considering our first chosen aspect, Paul's mention of Adam requires our attention. Our interest is not only in whether Paul wrote of specifically physical death, but in whether he wrote about its origin. Paul states that sin and death "came into the world" and that they did so "through one man" whom he names as "Adam" (v. 14). This reference to the first-named human in the biblical account alerts us to the possibility that he shared our present interest in origins.

In 1 Corinthians 15:22, Paul presents Adam and Christ as figureheads representing two overlapping eschatological categories of humanity: that "in Adam" and that "in Christ." In Romans too, Adam is doubtless a representative figure, the summation of an era brought to closure and replacement by Christ. This is perhaps clearest in verses 18 and 19, each of which sets a "one–all/many" against another. Some commentators emphasize this representative function of "Adam" to the exclusion of any sense of particularity and historicity. For Ziesler, Adam is indistinguishable from the human race, "a representative figure, Everyman and Everywoman, whose story is the story of all who share his sort of humanity."[8] Barth understands Adam as so much a dialectical foil for Christ that "he [Adam] exists only when he is dissolved, and he is affirmed only when in Christ he is brought to nought."[9] While recognizing Paul's employment of "Adam" as a representative of all and a foil for Christ, we, however, wish to hold these alongside an evident particularization and historicization of Adam in Romans 5:12–21. In this brief paragraph, Paul refers eight times to Adam's "one" default but only three times to the "many trespasses" of all.[10] Paul gives Adam a particularity in Romans 5 distinct from the race he represents.

---

8. Ziesler, *Romans*, 149.

9. Barth, *Epistle*, 171.

10. The sin of the "one": vv. 12, 14, 15, 16 (twice), 17, 18, 19; the sin of the "many": vv. 12, 16, 19.

Furthermore, Paul presents Adam not just as an individual but as a historicized individual. Without going as far as to deduce that Paul thought Adam to be historical in the same sense as Jesus—"an act in mythic history can be paralleled to an act in living history without the point of comparison being lost"[11]—Paul's portrayal of Adam contains an inseparable time element. The Adam–Christ antithesis bears an implicit before-and-after, with one "act of righteousness" (v. 18) forming the critical point of transition but with which other events are in chronological series, notably the coming of the law (vv. 13–14, 20) and, before that, a significant first trespass. Moule overstates this retrospective aspect of the passage when he suggests that "Paul reaches back to the beginnings of things . . . he re-writes the creation narrative."[12] Rather, Kreitzer's assessment is more accurate: "the protological dimension of the Adam/Christ analogy does come to the fore [in Romans 5] in a way that it does not in 1 Corinthians 15."[13] We can therefore expect it to be possible to make inferences from Romans 5 as to Paul's view of the origin of human death.

The Relation of Death with Sin

Paul brings "death" and "sin" into close collocation repeatedly in chapters 5–8 (5:12, 21; 6:16, 23; 7:5 cf. 7:11, 13; 8:2). He is not the first canonical author to express a relation between sin and death, but he is the first to present the relation in generic terms, writing as he does of their entering "the world" of human existence (vv. 12–13). As we saw in our survey of OT perspectives on death, its link there with sin was mainly limited to capital punishment.[14] It was not until the intertestamental period that a generic linking of sin and death emerged.[15] Of particular note in connection with

11. Dunn, *Romans*, 289. Hence our assertion that, for Paul, Adam is historicized, not historical.

12. H. C. G. Moule, quoted in Drummond, "Romans 5:12–21," 67–69.

13. Kreitzer, "Adam," 13.

14. See chapter 4, "Death in the Old Testament."

15. See chapter 4, "Death in the Intertestamental Literature."

Romans 5:12–21 is the depiction in the intertestamental literature of an enduring link between sin and death that was first forged by Adam. For example, 2 Baruch 54:15:

> For although Adam sinned and has brought death upon all who were not in his own time, yet each of them who has been born from him has prepared for himself the coming torment.[16]

The juxtaposition of "sin," "death," and an Adamic motif in Romans 5:12–21 suggests that Paul was himself familiar with this tradition, and the brevity with which he incorporates it into his letter—"Just as . . . "—indicates that (here as in 1 Corinthians 15) he assumed familiarity with the tradition amongst his readers. It was a tradition with a conception of death that most probably included human physical mortality, as noted earlier.[17]

The likelihood that the sin–death collocation in Romans is not a Pauline *novum* does not, however, relieve us of the need to examine the implicit nature of the link between sin and death in his letter, particularly when we are in pursuit of specifically Paul's meaning for "death." Two possibilities for a connection between sin and "death" are apparent: that "death" is a divine punishment for sin, or that it comes as a natural consequence of sin. Discussion of which of these Paul placed greatest emphasis on has centered around the sense in which he announces, in the first chapter of Romans, that "the wrath of God is [being] revealed" (1:18).[18] In our view, it is difficult to maintain from Romans that Paul held death (in any sense) to be divine retribution for sin. Admittedly, in 1:32 Paul alludes to the penalty of physical death for acts that carried it in the Levitical legislation (Leviticus 20:13), but the most he could have meant is that such Gentiles are *deserving* of death, not that they do die on account of their sin. Paul's declaration of the revealing of God's wrath (1:18) is followed not by a listing of the

---

16. See also 2 Baruch 23:4; 4 Ezra 3:21–26; 7:118–19.

17. See chapter 4, "Death in the Intertestamental Literature."

18. The debate was precipitated by the publication of C. H. Dodd's *The Epistle of Paul to the Romans* (1932).

forms that wrath now takes in the facts of human existence but by a listing of acts of sin that warrant God's wrath, to be expressed at a future judgement on a "day of wrath" (2:5). Presently, even while humankind is mortal (1:23), God leaves sin *un*punished (3:25 cf. 9:22). In our estimation, Barth's supposition that "in its [death's] inevitability we are reminded of the wrath which hangs over the man of the world and the world of man"[19] is not a fair statement of Paul's conception of death in relation to sin. Rather, "through the law," not death, "comes the knowledge of sin" (3:20 cf. 4:15; 7:7).

The other conceivable way in which Paul may have understood death to follow from sin is by natural consequence, such that the seeds of the sinner's demise are sown in their acts of sinning. The ethical passages in his letter could point towards Paul's having thought this: "sin . . . leads to death" (as a consequence) just as "obedience . . . leads to [i.e., develops rather than deserves] righteousness" (6:16). Sin pays its own wage of "death," whereas God gives "eternal life" (6:23). An indication of how, for Paul, sin might bring on "death" as a natural consequence—even physical death—could be thought to lie in Paul's depiction of sin as essentially carnal. Throughout Romans 1–8, Paul portrays the body as the locus of sin. In his description of "all ungodliness and wickedness" (1:18) he moves quickly from mention of "futile thinking . . . senseless minds . . . [and] lustful hearts" (1:21, 24a) to the bodily expression of those (1:24b, 26–27). He reasserts the carnality of sin frequently in chapters 6–8 (6:6, 12–13, 19; 7:5, 23; 8:13), not least in his use of *sarx* ("flesh") to denote the sphere of human living that is in opposition to life "according to the Spirit" (7:5, 18, 25; 8:3–9, 12–13). One could argue from this clear picture of sin bodily expressed that bodily death lay within Paul's purview when he connected the origins of sin and "death" so closely in 5:12. However, a physicality for sin does not necessarily imply physical death. Paul wrote of a currently living and breathing body (his own?) being "of death" (7:24) and "dead" (8:10); for him, even bodily "death" can be metaphorical, a qualitative diminution of life, not its quantitative end.

19. Barth, *Epistle*, 167. Note that Barth has here discarded his earlier nuanced conception of "death" for a singularly physical one.

This last observation leads us to suspect that pondering whether Paul thought death a punishment for or consequence of sin is a wrong way of framing the options. His assertion that "sin came into the world . . . and death came through sin" (5:12) does, we concede, couch "sin" and "death" as a sequential pair, but Paul may have conceived of them as more profoundly concomitant. The difficulty we encounter in identifying the relation as one of punishment or consequence supports Black's assessment that "Paul never reconciles the tension between death as arising from sin and as punishment for it. He shows no interest in doing so."[20] Perhaps Paul judged sin and "death" to interlock too closely for any attempt to articulate how one is the corollary of the other to be conclusive. Our inconclusive search for a chronological relation between them (as punishment or consequence) indicates it unlikely that Paul meant by "death" strictly physical death, because such death is necessarily chronologically subsequent to sin.[21]

A Continuum of Death Due to a Continuum of Sin

Earlier, we observed that although, for Paul, Adam is a representative of all, he stands out as a particular one in distinction from the rest of his race.[22] He is particular by way of his founding, initiating role. Also, we have argued that, in Paul's conception, "sin" and "death" are a synchronous pair. We are thus led to the question of the relation in Paul's mind between, on the one hand, Adam's particular sin and death, and, on the other, the sinning and dying that is endemic in our race. This book is motivated by a desire to know the origin of the human physical mortality that we currently observe, and our search rests on an assumption of a continuum of such mortality since its point of origin. Notwithstanding our still

20. Black, "Perspectives," 430.

21. To limit the options to punishment or consequence is to presuppose as physical the sense in which Paul used "death," and so to foreclose the question at hand.

22. See the section "'Adam' in Romans 5," above.

open question of whether specifically *physical* mortality belonged within Paul's concern, it is clear that he too assumed continuity between "death" at its origin and as a contemporary reality. After "death came" it "spread to all" (5:12); it was a reality over the generations "from Adam to Moses" (v. 14) and many more since (v. 15?). Our assumption of continuity being shared by Paul, we are led further to inquire whether Paul's words also explain how "death" has been perpetuated down to the present. This may help us to establish whether "death" for Paul included a physical sense.

Even though Paul's emphasis in 5:12a is on what happened "through one man," "death" implicitly entered "the world," a category of existence larger than that of a single individual. Indeed, "death" in Romans 5 is never that of an individual, even Adam; it is in every mention a phenomenon of universal reach, though manifest in each of "the many" (v. 15). It is, in contrast, sin to which Paul gives particularity, in the one man's transgression (vv. 14–19), although it too entered "the world" larger than Adam (vv. 12–13). In view of his collocation of death and sin, any inquiry into Paul's understanding of the way "death" was perpetuated from Adam onwards must focus on the role of sin. More precisely, it must focus on the particular sin of Adam and the universal sinning of all, and cannot avoid pondering the relation between these two.

Since Paul opens 5:12 with mention of sin coming "through one man," it is Adam's sin that is primarily implicated in any etiology of "death" here. This coming of sin created an opportunity for "death" to appear "through sin." The explicit bridge in 5:12 from this founding event to its subsequent universal legacy is "and so": "and so death spread to all." It is not possible to escape the conclusion, then, that in Paul's mind an ancestral sin initiated the thereafter widespread phenomenon of "death."

This is not to say that he thought "death" a wholly inexorable phenomenon in which each person plays no part, for his final phrase in 5:12, *"eph ho* all sinned" (v. 12d), may implicate every individual in his/her own "death." The crux of the matter is the meaning of *eph ho*. It would take us away from our main focus to consider each of the numerous suggestions of English translations.

Together, they amount to a soup of finely nuanced prepositions and pronouns. Suffice it to say that the suggestions can be divided into those which couch *eph ho* as a relative pronoun (e.g., "in whom") and those which take it as a conjunction (e.g., "because"). This division roughly correlates with understanding verse 12d as relating the sin of "all" ("all sinned") either to the sin of Adam (mentioned at the opening of the verse) or to the death of all (in the previous clause). In the first option, the emphasis is on Adam's culpability for our sin and so also for our dying. We have already been able to conclude this from verse 12a–c.[23] The second option carries the implication that we are all culpable; we, like Adam, sin, so we "die." If verse 12d says anything new—and it is unlikely to be redundant in a paragraph so tightly written—then it must be this. It seems to us, then, that Paul thought in terms of a "dual causality"[24] of "death": Adam's sin, *and* the sinning of everyone since. The second of these would not be a reality without the first, but it cannot be subsumed under the first.

Dunn presumes that the two "causes" of "death" reflect a distinction for Paul between two kinds of death: "natural" (a carry-over from Adam; physical mortality) and "spiritual" (brought on ourselves; moral degeneracy).[25] It is difficult to avoid drawing this distinction, although the fact that Paul never expresses it suggests that for him it was not a sharp one. Its being blurred prevents us from excluding physical mortality from Paul's conception of "death" in 5:12. Indeed, when Paul continues, with mention of "death . . . from Adam to Moses" (v. 14), it is probable that physical demise was uppermost in his mind, since the generations between Adam and Moses were, at the time of Paul's writing, patently dead and gone whereas their spiritual state of "life" or "death" was no longer plain for him or his readers.

23. See previous paragraph.
24. Fitzmyer, *Romans*, 407.
25. Dunn, *Paul*, 96.

Death Personified

Barth brings together the collocations of "sin" and "death" that occur at the beginning (v. 12) and end (v. 21) of our passage of primary interest in a paraphrase that expresses well the tight nexus between these two in Paul's conception: "Where sin lives, death lives in sin—and we are not alive. Where sin reigns, it reigns in death—and we are dead."[26] This paraphrase points up another feature of Paul's portrayal of "death" in Romans: his personification of it. "Death" is an agency that "came" (v. 12) and "reigned" (vv. 14, 17) but whose "dominion" no longer applies to Christ (6:9). Likewise for "sin" (5:12, 21; 6:12), which also enslaves (6:6, 14, 16–20) and deceives (7:11).

It is not easy to know how much to make of this ornate depiction. Some commentators take Paul's language as a cue to embark on flights of fancy, supposing Paul to have thought "sin" (and presumably "death" too) to be "not a human disposition or flaw in human nature, but an upper-case Power"[27] that "strode upon the stage of human history"[28] and "once set loose, cannot easily be stopped."[29] Another thinks Paul more prosaic, concerned "not so much to designate them [sin and death] as cosmic powers as to characterize them as forces of existential reality."[30] We judge, from Paul's ability to write elsewhere in his letter of both sin and death in plainer terms (e.g., 2:12; 3:23; 4:19; 5:7) that his personification of them in our passage of interest shows that he thought of them as more than mere existential realities. At the same time, we note that Paul, in portraying "death" as a power, is more restrained than the writer of Wisdom of Solomon before him: Paul avoids mention of the devil in Romans 5:12, a verse otherwise closely similar to Wisdom 2:24a, "through the devil's envy death entered the world." This restraint serves to support de Boer's contention that "the fact

---

26. Barth, *Epistle*, 169.
27. Gaventa, "Power," 231.
28. Fitzmyer, *Romans*, 411.
29. Ziesler, *Romans*, 145.
30. Dunn, *Romans*, 272.

that Paul speaks of death rather than Satan or the devil indicates that the cosmological power of death is strictly correlated with its anthropological effects."[31]

It is difficult to imagine that physical mortality was not, for Paul, at least one of these effects, because it is such a large fact of anthropological life as we know it. We can speculate further that, if so, Paul would have held one of two views: either that there was a time before the cosmological agency of "death" took effect, when a pristine primal human(ity) was, accordingly, physically immortal; or that physical mortality was a pre-established feature of human existence before it got co-opted by the power that now wields it, taking on an additional, epiphenomenal dimension simultaneous with sin. But on these alternatives we do well to be as reticent as Paul himself.

"Death" as the Absence of "Life"

There is no consensus as to what the "therefore" beginning Romans 5:12 is there for, except that it acts as a conjunction between this verse and the content of the letter up until this verse. The conjunction may operate on several levels, conceivably one of which is to continue and develop the distinction between "life" and "death" introduced just two verses previously (5:10). A distinction thus made, Paul, in our passage of primary interest and indeed through much of chapters 6–8, "sets off the gloom of death against the brighter beams of life and salvation."[32] In these chapters, *zōē* ("life") and its cognates appear on only a few occasions to denote natural, physical existence (see 7:1–2, 9; 8:12–13, 38). Much more frequently, "life" carries a "superabundant sense"[33] indicated by its occurring three times in the bound phrase *zōē aiōnios* ("eternal life"; 5:21; 6:22–23) and its association with "righteousness" (5:17–18 cf. 8:10).

31. Boer, *Defeat of Death*, 183.
32. Thomas, "Meaning," 208.
33. Scott, "Life," 555.

The full elucidation of what Paul meant by "life" in this sense, if possible at all, is beyond our present scope, but it seems clear to us that he meant more—much more—than the enlivening of our physical bodies, and that (by inference) his meaning for "death" was not confined to our physical demise. Suffice it to list three reasons for thinking this. Firstly, since righteousness and "life" come in tandem (as do sin and "death") "through one man" (5:17–18), "life" is a soteriological status, not an anthropological state. Secondly, "newness of life" is the goal of Paul's ethical appeal to those who have been "baptiz[ed] into death" while yet physically mortal (6:4, 11). Thirdly, correct as Fitzmyer may be that "the adjective 'eternal' [aiōnios] indicates the quality of that life rather than its duration,"[34] eternal life either outlasts physical life or else is a misnomer—but physical life, for Christian and non-Christian alike, is clearly brief. On all three counts, "life" does not approximate to its physical kind, so neither does its opposite, "death."

Related to the second and third points just made is the likelihood that "[eternal] life" is a present reality, not only a future prospect. As Wright comments, it would certainly be an uncharacteristic "cruel tease" of Paul's had he not been sincere in his choice of the present tense to describe his recipients as "those who have been brought from death to life" (6:13).[35] But the future tense dominates. In some instances (6:5, 8) the future tense is conceivably only logical ("if . . . then now")[36] but in chapter 8 it is unmistakeably the real future, since there the "life" given is applied to our presently "mortal bodies" (8:11 cf. 8:23). Here Paul revisits the resurrection theme of the climactic fifteenth chapter of his earlier letter to the Corinthians, and we face the question of whether one can extrapolate backwards from an eschatological statement of future resurrected bodily immortality to a protological assumption of an original physical immortality. In our judgement, since Paul made no such inference, neither should we.

---

34. Fitzmyer, *Romans*, 421.

35. Wright, *Resurrection*, 251.

36. Ibid., 251.

If it be contested that Paul implied an original physical immortality by depicting "the whole creation" as having been "subjected to futility" and awaiting a liberation "from its bondage to decay" that will include "the redemption of our bodies" (8:21–23), we respond with the comment that "labor pains" (v. 22) is an image that looks forward, not backward, to a state of affairs that has never yet been. But a full consideration of an apocalyptic passage such as this, in which there is no explicit mention of death, would take us beyond our narrow concern with Paul's perspective on human physical demise.

Conclusion

In our examination of Romans 5:12–21, we first detected, in his use of "Adam" here, Paul's partial concern with origins. Then, looking at three aspects of his portrayal of "death," we found indications that he meant by this term more than the ending of our physical lives: (i) "death" is concomitant with, not subsequent to, sin; (ii) as an agency, "death" is less than a diabolical power but, in twin dominion with sin, it is not a mere physical phenomenon; (iii) "[eternal] life" looms larger than physical life—and physical death. "Death" in this inflated, Pauline sense has no rightful place in the world it entered, and one day will have no place at all.

Even though Paul's conception of "death" in Romans 5–8 reaches further than human physical mortality, it probably includes the phenomenon of physical mortality. Paul shows signs of having been conversant with the intertestamental Jewish tradition which held that human physical mortality had to do with an Adamic first sin. Also, his recollection of persons long physically dead ("from Adam to Moses") in connection with death's "dominion" (5:14) indicates a physical aspect also for the "death" that first "came" (5:12). Moreover, it must have been difficult for Paul, as it is for us, to write of human death in a metaphorical sense without the literal sense forcing itself on his consciousness. On these grounds, in our judgement Paul probably presumed an original physical immortality for humans. We claim only probability, not certainty, for this judgement, though, because it concerns a detail that contributes little to Paul's message; whatever the origin and dominion of sin and death, they are now overshadowed by the greater abundance and dominion of grace (5:17, 20–21).

7

Last Words

WE STARTED OUT IN chapter 1 by posing the question "Why do we die?" and setting two alternative answers against each other. On the one hand, in Christian history it has mostly been held that age-related physical mortality is an acquired characteristic of human being, contrary to the Creator's original intention for us. On the other hand, scientific study of hominin remains presents a picture of developmental continuity between our present-day physical form and that of prehuman species, such that human beings have always been as physically mortal as all other forms of biological life.

In order to decide between these two positions, it is first necessary to decide what counts as supporting evidence. The traditional Christian view rests on an interpretation of biblical texts. In our exegesis of the two passages of principal relevance, we found that the Eden narrative contains no etiology of death, as often supposed, but that Paul, in Romans 5, probably implied that physical mortality was not at first a feature of human exis-tence. There is, then, a legitimate biblical basis for the traditional Christian position, albeit reduced to an ancillary meaning in a single passage. But, as we argued in chapter 3, the fact of a biblical proposition on our topic does not establish the traditional view as correct, because the Bible is not inerrant in its propositions about phenomena in nature.

Scientific inquiry is not immune to error either, but it is difficult to conceive of any other interpretation of the fossil finds than
that human physical origins lie in other mortal forms of biological life. The span of antiquity of the fossils is beyond doubt, and
they exhibit a gradation in anatomical features of the range seen
between modern great apes and modern humans. These empirical observations give the deduction of the evolutionary origin of
our bodies the status of, if not natural fact, then not far short. In
the absence, so far, of the discovery of the physical basis of age-
related mortality,[1] the (near) fact of our evolutionary origin is the
only one raised in our discussion so far that can be brought to
bear on our question—and it is clear which of the two alternative
answers it supports.

An evolutionary origin for the human body is not, it must be
conceded, conclusive for the view that humans have always been
inherently physically mortal. It is, of course, conceivable that the
earliest human individual(s) was (were) an immortal exception to
their mortal non-human predecessors. Such a being is envisioned
by C. S. Lewis as one whose "organic processes obeyed the law of
his own will, not the law of nature" so that "the length of his life
was largely at his own discretion."[2] Indeed, the traditional Christian belief in an original human physical immortality could not
have arisen—nor could Paul have implied it—were it not conceivable. The biblical accounts of God acting in nature by miracle, and
above all the miraculous foundation of Christian belief (in the
incarnation and resurrection), require us to be open to the possibility of subversion of the normal processes of biology. But, though
our imaginations are wont to seize on this possibility, and though
we are free to build doctrinal castles in the air, we do best, so far
as the natural history of our bodies is concerned, to keep a tight
reign on our speculations. Even Lewis, after offering a bold recon-
struction of what lies behind "an impenetrable curtain,"[3] admits
that "when we talk of what might have happened, of contingencies

1. See chapter 2, "The Questions Science Can, and Cannot Yet, Answer."
2. Lewis, *Problem*, 62.
3. Ibid., 59.

outside the whole of actuality, we do not really know what we are talking about."[4] Physically mortal human existence is the only actuality we now know, and the paleoanthropological findings give us strong reason to think it an actuality that we have always shared with all other forms of biological life. If, in addition to the scientific evidence, biblical precedent is sought for this view of physical mortality as intrinsic to the human constitution, it is not difficult to find amongst the Bible's diversity of perspectives on death, especially in the OT.[5]

An Additional Consideration

In weighing the twin facts of (i) an implication from Paul and (ii) the fossil finds, we have just argued that only the second of these, not the first, counts in support of its corresponding view. But even if the argument for an original immortality can draw no valid support from the fact of a statement in Scripture, there is another fact to which its adherents take recourse: the fact of how many people *feel* about physical death. Indeed, in chapter 1 we presented the Christian assumption of an original immortality as stemming primarily from sentiment, and only secondarily from certain Bible passages. The way people feel about death is as much a fact of the world around us as are fossils in the sand, but on reflection it turns out to be a fact with as little bearing on the origin of biological death as the fact of a biblical author's utterance.

As the *Epic of Gilgamesh* shows, humans' awareness of their mortality is at least as old as civilized history.[6] It is safe to say that, over the millennia, humans have generally perceived their mortality negatively. Why this is the case is at once easy and very difficult to say. Tillich may have reached further than most in his claim that the prospect of our physically dying forces on us a realization of the

---

4. Ibid., 67.
5. See chapter 4.
6. See chapter 5, "Mesopotamian Myths as an Interpretive Aid."

possibility of our nonbeing.[7] This realization, among others, issues in unavoidable anxiety. Though unavoidable, people nevertheless attempt to avoid it. Tillich mentions the attempt to transpose the anxiety into lesser and more easily manageable fears,[8] but we can imagine a myriad of responses that could be grouped under the heading "denial of death." It is our contention that the traditional Christian assumption that physical mortality is not original to human being—especially as it now persists in the face of the scientific evidence—is in large measure a denial of death.

By accounting for physical dying in terms of its being an unnatural phenomenon consequent to sin, a Christian disowns his/her mortality, distances at least Christian others from their recent or imminent deaths, and disavows that God is responsible for anyone's physical end. It does not serve our purpose here to explore further the psychological function of the traditional belief; our claim is only that, rather than the widespread anxiety that surrounds death being an indication of the non-originality of human physical death, the belief that it is such an indication is itself an attempt to avoid the anxiety. Furthermore, even if humans had retained the physical immortality with which they were supposedly originally endowed, it is not self-evident that they would be free from anxiety, for our death-awareness is only (and only sometimes) an occasion for anxiety, not its source.

Implications of Our Conclusion

When a Christian worldview develops over a long period in the company of assumptions that later transpire to be false, some adjustment of that worldview is called for. Regarding the assumption that humans brought physical mortality upon themselves—an assumption that we have argued is false—two aspects of the Christian worldview are affected: the doctrine of the fall, and the understanding of the atonement.

7. Tillich, *Courage*, 44–52.
8. Ibid., 45.

The primary implication of intrinsic human physical mortality for understanding the fall is that the fall was of no direct consequence for the human body. The anatomical form and physiological functioning of the modern human frame is a carry-over from prehuman biological forms. So is its mortality. One may ask whether a gradualist scheme such as this can accommodate any conception of a historical fall at all. We believe it can, because the scheme retains an implicit non-human/human distinction. Exactly what this distinction consists in is difficult to define, but it would be generally agreed that it is, in part, a distinction between organisms that are not self-aware and we who uniquely are. An account of a fall can be built around the possibilities that self-awareness presents:

> As soon as a creature, man, became self-conscious and able to choose—and there must have been a real period of time, however imprecise, during which this transition occurred—then it became possible for actions and events to occur in the universe contrary to God's creative purpose.[9]

If the human body, as part of the material cosmos, was unaffected by "the free-willed failure of man to become what God intended,"[10] then another likely implication is that the rest of the cosmos was unaffected too. In other words, a strictly human, not cosmic, fall is most consistent with the outcome of our inquiry. We are bound therefore to reject depictions of nature as "tragic"[11] or "morbid or depraved"[12] as instances of the pathetic fallacy. Within our conception of a fall limited to the human will, we may allow that nature is "a good thing spoiled"[13] as long as the spoiling results from ongoing culpable human abuse[14] and not from an

---

9. Peacocke, *Creation*, 193.

10. Ibid., 153.

11. Tillich, *Shaking*, 87–89.

12. Lewis, *Miracles*, 124.

13. Ibid., 125.

14. So Bimson, "Fall," 75; Berry, "Earth," 45–46.

ontological event precipitated by a first human sin[15] or pre-human angelic sin.[16] If there is an irreducible residue of "natural evil" in the way the world works—and it is by no means clear to us that there is—then (i) we would wish to exclude age-related mortality from that category, and (ii) there would be urgent need for a theodicy that could accommodate such evil without recourse to a cosmic fall.[17] Both of these would be served by a measured notion of God's omnipotence and a less anthropocentric conception of God's goodness.

The atonement occupies a more central place in Christian belief than the fall, partly because it comprises the solution to a problem that is implicitly a real and present one, whatever its origin. Arguably, the atonement is *the* center of Christian belief. Its centrality means that we must consider the atonement in the light of our conclusion, but its profundity means that we cannot here provide more than a brief and inadequate sketch of the implications of our conclusion.

Of the several theories of the atonement, it is only those that hold the Christ-event to have been of objective effect that are relevant to our discussion, and, of those, only theories that locate the effect primarily *in* Jesus' crucifixion and partly *on* human mortality. Theories of atonement by way of the subjective influence of Jesus' ministry do not see his crucifixion as relevant to the ontological limits of human existence such as physical mortality.[18] Two theories of objective atonement include a stance on human physical mortality: *Christus Victor* and penal substitution. Between them, these theories have been dominant for most of Christian history.

The *Christus Victor* conception is of Jesus' death having brought about the defeat of Satan, to whom humanity was captive.

15. So Ham, "Death," 20.

16. So Alsford, "Evil," 128; Lloyd, "Humanity," 80–81; Lewis, *Problem*, 107–8.

17. For a review of attempts, see Southgate, "Theodicy," 808–15.

18. Note: In what follows, we assume the subjectivist view to be a partial, not complete, account of the atonement. We accept Gunton's cogent defense of an objective component in his *Actuality*, 155–60.

This was simultaneously the defeat of powers Satan wielded, notably the powers of sin and death: "In the death of Christ all the armed forces of evil were conquered: sin as the power of death, death as the power of the devil, and the devil as the power of sin and death."[19] Human physical mortality is almost certainly partly in question here, given the all-encompassing reach of this account,[20] although presentations of *Christus Victor* invariably omit to specify what "death" there denotes.

In penal substitution, Jesus, though himself undeserving of death, died the death that is otherwise the just desert of all fallen humans on account of their sin. Presentations of this conception are more often explicit in applying it to age-related mortality: "in his [Christ's] death, death itself, in its every expression, physical, spiritual and eternal, has its measure and mastery."[21]

In their full extent, these two theories of the atonement are very different (and have been developed somewhat in opposition to each other), but with regard to their perspective on human physical death they are quite similar. (i) They both presume physical mortality to be not inherent but to have entered human existence, differing only in their emphasis on the agent responsible: Satan (*Christus Victor*) or Adam (penal substitution). (ii) They both see human physical mortality as attendant on sin, either as a concurrent power (*Christus Victor*) or as acts of disobedience (penal substitution). (iii) They both draw an identity at some level between the death Jesus died and the age-related death we all die, such that he "died our death."[22]

We have already addressed the first two of these points in our handling of the biblical material: they are exegetically flawed[23] and hermeneutically outmoded.[24] As for the third point, while an

---

19. McDonald, *Concept*, 39.

20. Gunton highlights its "extension of the benefits of the divine victory to all parts of the created order" (Gunton, *Actuality*, 79-80).

21. McDonald, *Concept*, 40. See also Stott, *Cross*, 65.

22. Stott, *Cross*, 64. See also Lewis, *Miracles*, 134.

23. See chapter 5.

24. See chapter 3.

effectual atonement is indeed predicated on God having identified with humanity in the person of Jesus Christ—whether by representation or substitution, or both—we suggest it erroneous and unnecessary to conflate Jesus' dying and our physical mortality as equivalent in kind. It is erroneous because Jesus met a violent end by execution; his death was not simply a function of the physical mortality he (hypothetically) shared with the rest of humanity. It is unnecessary because within precisely those *exceptional* circumstances of Jesus' death there is surely scope for understanding its atoning significance independent of the general fact of human physical mortality. When Jesus' death is considered not in isolation but integral with his prior ministry and subsequent resurrection, the scope is greater still.

In recent years, expressions of objective atonement have been proffered that recognize the rich variety of associated NT metaphors, in addition to those of "victory" (*Christus Victor*) and "justification" (penal substitution).[25] Some of these are attempts to recontextualize NT soteriology for contemporary cultures.[26] Especially in the West, there remains a pressing need also to articulate a model of objective atonement in cognizance of the scientific indications of original human physical mortality. We hope that this book has helped to prepare the ground towards that end.

25. See: Green and Baker, *Scandal*, 35–115; Green, "Kaleidoscopic View," 157–85; Williams, "Unified Theory," 228–48.

26. See Green, *Scandal*, 153–221; Robbins, "Atonement," 329–44.

Bibliography

Alexander, Denis. *Rebuilding the Matrix: Science and Faith in the 21st Century.* Grand Rapids: Zondervan, 2003.

Alsford, Sally. "Evil in the Non-Human World." *Science and Christian Belief* 3 (1991) 119–30.

Bailey, Lloyd. *Biblical Perspectives on Death.* Philadelphia: Fortress, 1979.

———. "Death as a Theological Problem in the Old Testament." *Pastoral Psychology* 22 (1971) 20–32.

Barth, Karl. *The Epistle to the Romans.* Translated by Edwyn C. Hoskyns. Oxford: Oxford University Press, 1933.

Beattie, Derek R. G. "What Is Genesis 2–3 About?" *Expository Times* 92 (1980) 8–10.

Berry, Robert J. "This Cursed Earth: Is 'the Fall' Credible?" *Science and Christian Belief* 11 (1999) 29–49.

Bimson, John. "Reconsidering a 'Cosmic Fall.'" *Science and Christian Belief* 18 (2006) 63–81.

Black, C. Clifton. "Pauline Perspectives on Death in Romans 5–8." *Journal of Biblical Literature* 103 (1984) 413–33.

Blocher, Henri. *In the Beginning.* Leicester: InterVarsity, 1984.

Boer, Martinus C. de. *The Defeat of Death: Apocalyptic Eschatology in 1 Corinthians 15 and Romans 5.* Journal for the Study of the New Testament, Supplement Series 22. Sheffield: JSOT Press, 1988.

Clark, William R. *Sex and the Origins of Death.* Oxford: Oxford University Press, 1996.

Coats, George W., "The God of Death: Power and Obedience in the Primeval History." *Interpretation* 29 (1975) 227–39.

Dodd, C. H. *The Epistle of Paul to the Romans.* London: Hodder & Stoughton, 1932.

Drummond, Alistair. "Romans 5:12–21." *Interpretation* 57 (2003) 67–69.

Duce, Philip P. "Comment on 'This Cursed Earth.'" *Science and Christian Belief* 11 (1999) 161–63.

———. "Complementarity in Perspective." *Science and Christian Belief* 8 (1996) 145–55.

Dunn, James D. G. *Romans.* Word Bible Commentary 38A. Dallas: Word, 1988.

———. *The Theology of Paul the Apostle.* Edinburgh: T. & T. Clark, 1998.

Fitzmyer, Joseph A. *Romans*. Anchor Bible 33. New York: Doubleday, 1992.

Gaventa, Beverly R. "The Cosmic Power of Sin in Paul's Letter to the Romans." *Interpretation* 58 (2004) 229–40.

George, Andrew, translator. *The Epic of Gilgamesh: The Babylonian Epic Poem and Other Texts in Akkadian and Sumerian*. London: Penguin, 1999.

Gibson, John C. L. *Genesis*. Vol. 1. Daily Study Bible. Edinburgh: Saint Andrew, 1981.

Goldingay, John. *Models for the Interpretation of Scripture*. Toronto: Clements, 2004.

———. *Models for Scripture*. Carlisle: Paternoster, 1994.

Green, Joel B., and Baker, Mark D. *Recovering the Scandal of the Cross*. Downers Grove, IL: InterVarsity, 2000.

———. "Kaleidoscopic View." In *The Nature of the Atonement: Four Views*, edited by James Beilby and Paul R. Eddy, 157–85. Downers Grove, IL: InterVarsity, 2006.

Gunton, Colin. *The Actuality of Atonement*. Edinburgh: T. & T. Clark, 1988.

Ham, Ken. "Two Histories of Death." *Creation* 24 (2001) 18–20.

Hyers, Conrad. *The Meaning of Creation: Genesis and Modern Science*. Atlanta: John Knox, 1984.

Johnston, Philip S. *Shades of Sheol: Death and Afterlife in the Old Testament*. Leicester: InterVarsity, 2002.

Kreitzer, Larry J. "Adam and Christ." In *Dictionary of Paul and His Letters*, edited by Gerald F. Hawthorne, Ralph P. Martin, and Daniel G. Reid, 9–15. Leicester: InterVarsity, 1993.

Lewis, C. S. *Miracles*. London: Fount, 1960.

———. *The Problem of Pain*. London: Fontana, 1957.

Lloyd, Michael. "The Humanity of Fallenness." In *Grace and Truth in the Secular Age*, edited by Timothy Bradshaw, 66–82. Grand Rapids: Eerdmans, 1998.

Lockwood, Charles. *The Human Story: Where We Come from and How We Evolved*. Updated ed. London: Natural History Museum, 2013.

Lucas, Ernest. "Some Scientific Issues Related to the Understanding of Genesis 1–3." *Themelios* 12.2 (1987) 46–51.

Magalhães, João Pedro de. senescence.info 1997–2014. http://www.sene scence.info.

Martin-Achard, Robert. *From Death to Life: A Study of the Development of the Doctrine of the Resurrection in the Old Testament*. London: Oliver & Boyd, 1960.

McDonald, H. Dermot. *The New Testament Concept of the Atonement*. Cambridge: J. Clarke & Co., 2006.

McGrath, Alister. *Christian Theology: An Introduction*. Oxford: Blackwell, 1997.

Moberly, R. Walter L. "Did the Serpent Get It Right?" *Journal of Theological Studies* 39 (1988) 1–27.

Morris, Leon. *The Epistle to the Romans*. Pillar New Testament Commentary. Leicester: InterVarsity, 1988.

————. *The Wages of Sin: An Examination of the New Testament Teaching on Death*. London: Tyndale, 1955.

Peacocke, Arthur. *Creation and the World of Science*. Oxford: Clarendon, 1979.

————. *Science and the Christian Experiment*. Oxford: Oxford University Press, 1971.

Ramm, Bernard. *The Christian View of Science and Scripture*. Grand Rapids: Eerdmans, 1954.

Robbins, Anna M. "Atonement in Contemporary Culture: Christ, Symbolic Exchange, and Death." In *The Atonement Debate: Papers from the London Symposium on the Theology of Atonement*, edited by Derek Tidball, David Hilborn, and Justin Thacker, 329–44. Grand Rapids: Zondervan, 2008.

Schmithals, Walter. "Death, Kill, Sleep." In *New International Dictionary of New Testament Theology*, edited by Colin Brown, 429–47. Exeter: Paternoster, 1975.

Scott, J. Julius, Jr. "Life and Death." In *Dictionary of Paul and His Letters*, edited by Gerald F. Hawthorne, Ralph P. Martin, and Daniel G. Reid, 553–55. Leicester: InterVarsity, 1993.

Silberman, Lou H. "Death in the Hebrew Bible and Apocalyptic Literature." In *Perspectives on Death*, edited by Liston O. Mills, 13–32. Nashville: Abingdon, 1969.

Skinner, John. *Genesis*. International Critical Commentary. Edinburgh: T. & T. Clark, 1930.

Southgate, Christopher. "God and Evolutionary Evil: Theodicy in the Light of Darwinism." *Zygon* 37 (2002) 803–24.

Stott, John R. W. *The Cross of Christ*. Leicester: InterVarsity, 1986.

Thomas, Richard W. "The Meaning of the Terms 'Life' and 'Death' in the Fourth Gospel and in Paul." *Scottish Journal of Theology* 21 (1968) 199–212.

Tillich, Paul. *The Courage to Be*. London: Collins, 1962.

————. *The Shaking of the Foundations*. Harmondsworth: Penguin, 1966.

Wenham, Gordon. *Genesis 1–15*. Word Bible Commentary 1. Waco, TX: Word, 1987.

Westermann, Claus. *Genesis 1–11*. Translated by John J. Scullion. London: SPCK, 1984.

Williams, David T. "Towards a Unified Theory of the Atonement." In *The Atonement Debate: Papers from the London Symposium on the Theology of Atonement*, edited by Derek Tidball, David Hilborn, and Justin Thacker, 228–48. Grand Rapids: Zondervan, 2008.

Wolff, Hans W. *Anthropology of the Old Testament*. Philadelphia: Fortress, 1974.

Wright, N. T. *The New Testament and the People of God*. Vol. 1 of *Christian Origins and the Question of God*. Philadelphia: Fortress, 1992.

————. *The Resurrection of the Son of God*. Vol. 3 of *Christian Origins and the Question of God*. Minneapolis: Fortress, 2003.

Ziesler, John A. *Paul's Letter to the Romans*. TPI New Testament Commentaries. London: SCM, 1989.

Index

www.ingramcontent.com/pod-product-compliance
Lightning Source LLC
Chambersburg PA
CBHW071107090426
42737CB00013B/2520